The Roman Army
and the New Testament

The Roman Army and the New Testament

Christopher B. Zeichmann

LEXINGTON BOOKS/FORTRESS ACADEMIC
Lanham • Boulder • New York • London

Published by Lexington Books/Fortress Academic

Lexington Books is an imprint of The Rowman & Littlefield Publishing Group, Inc.
4501 Forbes Boulevard, Suite 200, Lanham, Maryland 20706
www.rowman.com

6 Tinworth Street, London SE11 5AL

British Library Cataloguing in Publication Information Available

Library of Congress Cataloging-in-Publication Data

Names: Zeichman, Christopher B., author.
Title: The Roman army and the New Testament / Christopher B. Zeichmann.
Description: Lanham : Fortress Academic, 2018. | Includes bibliographical
    references and indexes.
Iders: LCCN 2018032322 (print) | LCCN 2018033173 (ebook) | ISBN
    9781978704039 (electronic) | ISBN 9781978704022 (cloth : alk. paper)
Subjects: LCSH: Bible. New Testament—Criticism, interpretation, etc. |
    Military art and science in the Bible. | Rome—Army—Foreign
    service—Palestine. | Rome—History—Empire, 30 B.C.-284 A.D.
Classification: LCC BS2361.3 (ebook) | LCC BS2361.3 .Z45 2018 (print) | DDC
    225.8/35500937—dc23
LC record available at https://lccn.loc.gov/2018032322

∞™ The paper used in this publication meets the minimum requirements of American
National Standard for Information Sciences—Permanence of Paper for Printed Library
Materials, ANSI/NISO Z39.48-1992.

Printed in the United States of America

To Johann Zeichmann, known most of his life
as the eldest John Zeichman.
Were it not for his refusal to submit to the draft,
the Zeichman family would not exist.

Contents

List of Figures   ix

Acknowledgments   xi

Abbreviations   xiii

Introduction: A War Machine in the Holy Land?   xvii

1   Who Were the Soldiers in Palestine?   1

2   What Did the Military Do in Early Roman Palestine?   23

3   The Military in the Gospels and Acts   49

4   The Military in the Pauline Corpus   107

5   The Military in Revelation   125

Conclusion: Reading a Complicated Bible in Complicated Times   139

Bibliography   145

Index   165

Author Index   169

Index of the Bible and Other Ancient Texts   173

About the Author   185

List of Figures

1.1 Eastern Portion of the Roman Empire, circa 44 CE.   3

1.2 Roman Military Diploma.   11

1.3 Plundering the Jerusalem Temple.   13

2.1 Roman Military Sites in Palestine, 63 BCE–132 CE.   30

2.2 Hadrian's Aqueduct in Judaea.   35

3.1 Neptune on an Inscription of *legio X Fretensis*.   51

3.2 Pig and Warship on a Brick Stamp of *legio X Fretensis*.   51

3.3 Claudius and Britannia.   69

3.4 Antonia Fortress.   87

4.1 Roman Imperial Triumph.   111

Acknowledgments

Labor, whether intellectual or otherwise, never occurs in a vacuum. Leif Vaage has been foremost among my conversation partners and his comments on this work have been pivotal. Several Toronto-based scholars have helped in various ways, including Terry Donaldson, John Kloppenborg, Scott Lewis, Colleen Shantz, and Natalie Wigg-Stevenson. Their conversations, support, and feedback are deeply appreciated. Though in the tundra of Regina, I must also thank William Arnal for his extensive assistance as well.

Others at Toronto also deserve to be mentioned, especially John Egger, who has been a loyal friend and dialogue partner since my first days as a doctoral student. Kieren Williams and Trent Voth deserve thanks for their conversations on the following matters. Jordan Balint and Dianne Everitt also read a full draft of this book and provided extensive feedback. Their generosity in taking time to read rougher versions of what is found here is a sign of true friendship and kindness.

The insights of my editor at Lexington/Fortress Academic—Neil Elliott—are too numerous to list here. I have long admired his scholarship and it is an honor to work with him. His support for this project is deeply appreciated and, as one raised in the Evangelical Lutheran Church in America, I find it satisfying to see this book published under the Fortress Academic imprint, an imprint with a Lutheran heritage.

Several others aided the creation of the *Database of Military Inscriptions and Papyri of Early Roman Palestine* (*DMIPERP*; see www.ArmyOfRomanPalestine.com), including William Arnal, Christian Remington, Erin Roberts, and James and Kathy Zeichman. Their help in this project has been invaluable, as it has rendered publicly accessible work on the military and the New Testament. As will become clear, *DMIPERP* forms an extension of this monograph, collecting otherwise-obscure texts that inform the interpretation of the history of early Roman Palestine and the New Testament. Down south, Laura, Jim, Colleen, and my parents have been helpful in too many ways to name here.

Finally, it is necessary to thank my occasional writing partner, Nathanael Paul Romero. Nathanael has been singularly helpful in thinking about state violence as a theoretical problematic needing redress, particularly vis-à-vis mundane practices. He has helped me understand state violence as a matter that extends beyond physical violence to many other aspects of the state. I end these acknowledgments with a quotation from Nathanael that also summarizes the orientation in the following pages: "My own politics are less centered on the extraordinary violence of so-called 'extremists' and those positioned as threats to 'our way of life' and more on the everyday violences that function to maintain that way of life."

Abbreviations

Citations from ancient literature follow *SBL Handbook of Style*, 2nd ed. (Atlanta: Scholars Press, 2014).

§ or *DMIPERP*	*Database of Military Inscriptions and Papyri of Early Roman Palestine* [www.ArmyOfRomanPalestine.com]
AE	*L'Année épigraphique*
ANRW	Hildegard Temporini and Wolfgang Haase, eds. *Aufstieg und Niedergang der römischen Welt: Geschichte und Kultur Roms im Spiegel der neueren Forschung.* Part 2. New York: de Gruyter, 1972–.
ASP	American Studies in Papyrology
BARBS	British Archaeology Reports, British Series
BARIS	British Archaeology Reports, International Series
BCAW	Blackwell Companions to the Ancient World
BGU	*Aegyptische Urkunden aus den Königlichen Staatlichen Museen zu Berlin, Griechische Urkunden.* 15 vols. Berlin: Weidmannische Buchhandlung, 1895–1983.
CBQ	*Catholic Biblical Quarterly*
CIL	*Corpus Inscriptionum Latinarum*
CPL	*Corpus Papyrorum Latinarum*
IG	*Inscriptiones Graecae.* 1873–.
IGR	René Cagnat, Jules Toutain, Georges Lafaye, and Victor Henry, eds. *Inscriptiones Graecae ad Res Romanas Pertinentes.* 4 vols. Paris: Leroux, 1911–1927.
ILS	Hermann Dessau, ed. *Inscriptiones Latinae Selectae.* 3 vols. Berlin: Weidmann, 1892–1916.
JBL	*Journal of Biblical Literature*

JRASup	Journal of Roman Archaeology Supplements
JRS	*Journal of Roman Studies*
JSHJ	*Journal for the Study of the Historical Jesus*
JSNT	*Journal for the Study of the New Testament*
JSNTSup	Journal for the Study of the New Testament Supplement Series
Mas	Hannah Cotton, and J. Geiger. *Masada II, The Yigal Yadin Excavations 1963–1965, Final Reports: The Latin and Greek Documents*. Jerusalem: Israel Exploration Society, 1989.
NewDocs	G. H. R. Horsley and S. R. Llewelyn, eds. *New Documents Illustrating Early Christianity*. Grand Rapids: Eerdmans, 1981–.
O.Ber.	*Documents from Berenike*. 2000–.
O.Amst.	Roger S. Bagnall, P. J. Sijpesteijn, and K. A. Worp, eds. *Ostraka in Amsterdam Collections*. Zutphen: Terra, 1976.
OGIS	W. Dittenberger, ed. *Orientis graeci inscriptiones selectae*. 2 vols. Leipzig: Hirzel, 1903–1905.
P.Amh.	B. P. Grenfell and A. S. Hunt, eds. *The Amherst Papyri, Being an Account of the Greek Papyri in the Collection of the Right Hon. Lord Amherst of Hackney, F.S.A. at Didlington Hall, Norfolk*. 2 vols. London: Oxford University Press, 1900–1901.
P.Corn.	William L. Westermann and Casper J. Kraemer, Jr., eds. *Greek Papyri in the Library of Cornell University*. 55 vols. New York: Columbia University Press, 1926.
P.Dura	Charles Bradford Welles, Robert. O. Fink, and J. F. Gilliam, eds. *The Excavations at Dura-Europos conducted by Yale University and the French Academy of Inscriptions and Letters, Final Report V, Part I, The Parchments and Papyri*. 155 vols. New Haven: Yale University Press, 1959.
P.Fouad	André Bataille, Octave Guéraud, Pierre Jouguet, Naphtali Lewis, H. Marrou, Jean Scherer, and W. G. Waddell, eds. *Les Papyrus Fouad I*. Cairo: Institut français d'archéologie orientale, 1939.
P.Gen.	*Les Papyrus de Genève*. Amsterdam: Hakkert, 1896–.
P.Lond.	*Greek Papyri in the British Museum*. 6 vols. London: British Museum, 1893–1924.
P.Mich.	*Michigan Papyri*. 1931–.
P.Oxy.	*The Oxyrhynchus Papyri*. London: Egypt Exploration Fund, 1898–.
P.Strassb.	Friedrich Preisigke, ed. *Griechische Papyrus der Kaiserlichen Universitäts- und Landes-bibliothek zu Strassburg*. Leipzig: Urkunden, 1912–.
P.Tebt.	*The Tebtunis Papyri*. 5 vols. London, 1902–2005.
P.Yadin	Yigael Yadin and J. C. Greenfield, eds. *The Documents from the Bar Kokhba Period in the Cave of the Letters: Aramaic and Nabatean Signatures and Subscriptions*. Judean Desert Studies 2. Jerusalem: Israel Exploration Society, 1989.

PSI	*Papiri greci e latini: Pubblicazioni della Società Italiana per la ricerca dei papiri greci e latini in Egitto.* 1912–.
RGZM	Barbara Pferdehirt, ed. *Römische Militärdiplome und Entlassungs urkunden in der Sammlung des Römisch-Germanischen Zentralmuseums.* 2 vols. Mainz: Römisch-Germanisches Zentralmuseum, 2004.
RMD	Margaret Roxan and Paul Holder, eds. *Roman Military Diplomas.* 1954–.
RMR	Robert O. Fink, ed. *Roman Military Records on Papyrus.* ASP 26. Cleveland: Press of Case Western University, 1971.
SB	*Sammelbuch griechischer Urkunden aus Aegypten.* 1915–.
SBLECL	Society of Biblical Literature Early Christian Literature
SBLSS	Society of Biblical Literature Seminar Series
SCI	*Scripta Classica Israelica*
SEG	*Supplementum Epigraphicum Graecum*
SemeiaSt	Semeia Studies
SJLA	Studies in Judaism in Late Antiquity
SNTSMS	Society of New Testament Studies Monograph Series
T.Vindol.	Alan K. Bowman, J. D. Thomas, and Roger S. O. Tomlin, eds. *Vindolanda: The Latin Writing Tablets.* 4 vols. 1983 2011.
TSAJ	Texts and Studies in Ancient Judaism
WUNT	Wissenschaftliche Untersuchungen zum Neuen Testament
ZPE	*Zeitschrift für Papyrologie und Epigraphik*

Introduction

A War Machine in the Holy Land?

The sky darkens as clouds obstruct the sun's light. Thunder rolls. Dozens watch as rain soaks the cloaks of friends, family, and foes beholding the emotional site at Calvary. The choir's reverent harmony crescendos as the camera stoically sustains a shot of Jesus' crucified body. Finally, we are presented with the image of a man of great dignity in military dress watching the proceedings closer than anyone else. This man, a Roman centurion, is played by none other than John Wayne—The Duke himself!—in a cameo appearance near the height of his celebrity. "Truly this man was the son of God," the centurion mumbles in a monotonous voice suggesting a state of mind somewhere between boredom and inebriation. The meaning of the centurion's confession at the cross is debated among biblical scholars: is it veneration ("I now understand that Jesus is the son of God!"), sarcasm ("Some son of God this guy was . . ."), or agony ("Jesus, not the emperor, is the son of God!")? The intonation of this pronouncement is paramount to its meaning. In whatever way the centurion said this line, one scholar joked, we can be certain it wasn't the manner in which John Wayne bludgeoned this iconic passage in *The Greatest Story Ever Told* (1965).

The military occupies a strange place in the study of the New Testament. It is both everywhere and nowhere, it is a matter of consensus and never-ending debate. No one doubts that it was a profoundly negative force in the lives of ancient Palestinians, but what does one make of the seemingly positive interactions between military men and early Christians? What did the centurion at the cross mean when he said Jesus was "(a) God's son"? Did soldiers abandon their ill treatment of Jewish civilians as John the Baptist demanded (Luke 3:14), or continue in their ways as usual? What does one make of the demon named "Legion" in Mark 5:1–20—is it a veiled criticism of Roman imperialism? How might one characterize the significance of the military in the New Testament more broadly? On the one hand, a scholar with no less pedigree than E. P. Sanders confidently dismisses its relevance since Rome did not rule Galilee—or Palestine more broadly—"by occupation."[1] Insofar as no legions garrisoned the region before the Jewish War (66–73 CE), Rome's methods of rule are to be found elsewhere: foremost in the proxy rule of client kings and the implicit assurance of Roman might. On the other hand, a growing number of scholars interested in anti-imperial and post-colonial New Testament interpretation maintain that the military was a major aspect of daily life in prewar Palestine. Representative of this more recent trend is Richard Horsley:

> The "forceful suasion" of Roman military power functioned through the perceptions of subject peoples. The Romans simply terrorized peoples into submission and, they hoped, submissiveness through the ruthless devastation of the land and towns, slaughter and enslavement of the people, and crucifixion of people along the roadways or in public places.[2]

With esteemed scholars diametrically opposed on the military's role in early Roman Palestine, it is little surprise that commentators often avoid the topic completely. Indeed, no monograph details the social history of the armies located in Palestine during the New Testament period. To be sure, there are two helpful volumes with sections devoted to Herod the Great's army, though neither book discusses the New Testament or the military of the first century CE, nor have they exerted much influence on New Testament scholarship.[3] The most frequently cited article on the topic, an *ANRW* entry by Roman military historian Denis B. Saddington titled "Roman Military and Administrative Personnel in the New Testament," evinces the author's inexperience with New Testament scholarship and is strikingly naïve when it comes to questions of historicity; Saddington states, for instance, that the Gospel of Peter and the Acts of Pilate may contain the names of the soldiers who crucified the historical Jesus—one would be hard pressed to find a biblical scholar so confident in their reliability.[4] It is becoming less necessary for scholars to rely upon Saddington's tenuous grasp of New Testament studies, thanks to two recent books on the literary depiction of soldiers in Luke-Acts and another book on the Jewish War and the Gospel of Mark, but their influence remains to be seen.[5] Instead, commentators more commonly rely on general studies of the Roman army that do not address the specifically *Palestinian* experience of the military, or adopt credulous readings of Josephus' wartime accounts and generalize on the basis of exceptional events. Despite the prevalence of military men and metaphors throughout the New Testament, there is a resounding silence from scholars on the topic.

The problem becomes even more complicated when one turns to the text of the New Testament itself. The narratives of the New Testament on the one hand present soldiers at the cross as Jesus' abusers, who mock, beat, and execute him (Mark 15:16–32). Yet the same collection of texts depicts the centurion Cornelius as first Gentile Christian (Acts 10) and praise an unnamed centurion for showing greater faith than all of Israel (Luke 7:1–10/Matt 8:5–13). Though there have been efforts to resolve these tensions, I will argue over the course of this book that the military was a sufficiently complicated issue in antiquity that it may be best not to downplay the ambivalence of New Testament texts toward the military, but to explore how different authors relate to and depict this institution.

The present book is both social-historical and literary in its methods, operating under the assumption that New Testament literature is elucidated by the social context in which it was produced. Chapter 1 asks the deceptively simple question, "who served as soldiers in Palestine?" How one answers this question will depend on several factors: at what time, where within Palestine, and what his rank was. In the prewar period, soldiers tended to be Palestinian locals (both Jews and Gentiles), though after the war they were almost exclusively Gentiles recruited from other provinces with a strong Roman identification. The findings from this chapter will allow us to assess the type of soldiers with whom civilians like the evangelists were interacting, and how they may have represented soldiers from the time of Jesus. Chapter 2 will address the next logical topic and discuss the military's activities in Palestine. Contrary to the popular image, warfare was a very small part of the soldiering life.[6] While the army was certainly not a benign institution, soldiers' duties rarely involved physical violence. Soldiers performed numerous roles, ranging from policing to infrastructural construction to mediating local disputes. Here we will see the ways in which civilians interacted with soldiers and how civilians represented those interactions in documentary and literary sources.

Following these social-historical discussions, chapters 3 through 5 will examine the depiction of the military within biblical texts, attending to the distinctive voice of each author. After contextualizing the composition of a given New Testament author, each chapter will discuss in detail all major references to the military by that author. Here, the implications of chapters 1 and 2 for biblical interpretation will become clear. It must be emphasized that these are only *partial* readings of the books in question—many important questions are simply outside this study's purview. Finally, the book will close with a conclusion that briefly reflects upon the social, theoretical, and theological implications of the foregoing study, devoting special attention to the question of ambivalence; the New Testament offers both "yes" and "no" responses to the military—a tension that is often present within individual books of the bible as well. What might one make of this inconsistency?

A FEW MATTERS OF DEFINITION

I use the term "Palestine" to refer to the contiguous lands controlled by Herod the Great and/or his descendants at their fullest extent (i.e., Judaea, Galilee, Peraea, Batanaea, Gaza), as well as the entire Decapolis (including Damascus) and independent cities in the area like Ascalon—the boundaries of this definition are visualized in figure 2.1. The primary reason for defining Palestine thus is to distinguish lands marked by Herodian history from other nearby land (e.g., Syria, Nabataea, Emesa). This distinction corresponds closely with the region in which the Gospels depict Jesus' activity, omitting only Tyre and Sidon. The word "Judaea" will almost always refer to the (sub)province/kingdom of Judaea to avoid confusion with the geographic region of Judaea in southern Palestine. That is, I employ "Judaea" in the same register as Syria, Italia, and Egypt; it is not an informal designation like Samaria, Idumaea, or the Golan Heights.

Though this book is not written on the contentious Jew-Judaean debate, I employ both of these terms and distinguish them from one another. This debate pertains to the meaning of the Greek term *Ioudaios*, whether it is best translated in a religio-cultural sense (i.e., "Jew"), in a geographic-ethnic sense (i.e., "Judaean"), or perhaps some other way. "Judaean" is used in this book as an adjective or noun in distinction from other geographic and provincial identities like Batanaean, Syrian, Egyptian, and Thracian. Moreover, since Judaean denizens were diverse in their religious and ethnic identities, I will sometimes clarify the person discussed is a Gentile Judaean or a Jewish Judaean (or Jewish Galilean, Jewish Batanaean, Jewish Syrian, etc.). "Judaism" and its derivatives are employed as a loosely ethno-religious category in this book, as it is difficult to distinguish between Samaritans and Jews in the surviving record and it is often unclear how much a given individual cared about being perceived as "Jewish."

In general, though, the term "Jew" will denote people recognized as an *Ioudaios* by their peers; one might be identified as an *Ioudaios* for different reasons, including religious affiliations, ethnic kinship, knowledge of Jewish practices, etc. Jewishness is not understood as a stable identity, but as a variable set of identification practices whose importance an individual might emphasize or downplay as context warranted. For instance, Josephus' claim that Tiberius Julius Alexander had abandoned his Judaism tells us more about the Jewish identity practices that Josephus found significant than Tiberius' own self-identification (*A.J.* 20.100); Tiberius may well have understood himself to be a Jew, even if he did not behave in a manner commensurate with Josephus' expectations.

Finally, this book makes extensive use of the *Database of Military Inscriptions and Papyri of Early Roman Palestine* [ArmyOfRomanPalestine.com]. *DMIPERP* collects all known military inscriptions and papyri relating to Palestine 63 BCE–132 CE, both in original language and in English translation. Texts in this database are regularly cited in this book, as indicated by the section symbol; for instance, if one sees the citation of §25 in the text of this book, it refers to *DMIPERP* entry number 25, which can be easily found on the aforementioned website. Though many scholars may be more familiar with other standard corpora (e.g., *CIL*, *RMD*, *CIIP*), *DMIPERP* collects all relevant texts into a single resource for greater convenience, as it may be difficult to track down one or another inscription without access to a research library or an interlibrary loan system.

NOTES

1. E. P. Sanders, "Jesus' Galilee," in *Fair Play: Diversity and Conflicts in Early Christianity. Essays in Honour of Heikki Räisänen*, ed. Ismo Dunderberg, Christopher M. Tuckett, and Kari Syreeni, Novum Testamentum Supplements 103 (Köln: Brill, 2002), 10.

2. Richard A. Horsley, *Galilee: History, Politics, People* (Valley Forge: Trinity Press International, 1995), 116.

3. Samuel Rocca, *Herod's Judaea: A Mediterranean State in the Classical World*, TSAJ 122 (Tübingen: Mohr Siebeck, 2008), 133–196; Israel Shatzman, *The Armies of the Hasmonaeans and Herod: From Hellenistic to Roman Frameworks*, TSAJ 25 (Tübingen: Mohr Siebeck, 1991), 139–309. Cf. Samuel Rocca, *The Forts of Judaea 168 BC–73 AD: From the Maccabees to the Fall of Masada*, Fortress 65 (Oxford: Osprey, 2008); Samuel Rocca, *The Army of Herod the Great*, Men-at-Arms 443 (Oxford: Osprey, 2009).

4. Denis B. Saddington, "Roman Military and Administrative Personnel in the New Testament," *ANRW* II.26.3: 2414. In a different article, Saddington indicates unfamiliarity with New Testament textual criticism, taking too seriously minor variants where *hekatontarchos* reads *chiliarchos* in Matt 8:5–13 ("The Centurion in Matthew 8:5–13: Consideration of the Proposal of Theodore W. Jennings, Jr., and Tat-Siong Benny Liew," *JBL* 125 (2006): 140–141).

5. Luke-Acts: Laurena Ann Brink, *Soldiers in Luke-Acts: Engaging, Contradicting and Transcending the Stereotypes*, WUNT II 362 (Tübingen: Mohr Siebeck, 2014); Alexander Kyrychenko, *The Roman Army and the Expansion of the Gospel: The Role of the Centurion in Luke-Acts*, Beihefte zur Zeitschrift für die neutestamentliche Wissenschaft 203 (Berlin: De Gruyter, 2014). Mark: Gabriella Gelardini, *Christus Militans: Studien zur politisch-militärischen Semantik im Markusevangelium vor dem Hintergrund des ersten jüdisch-römischen Krieges*, Novum Testamentum Supplements 165 (Leiden: Brill, 2016).

6. Those hoping for prolonged analysis of the various wars in Palestine during the early Roman period—however interesting their study might be—will find little discussion here, if simply to avoid an unwieldy scope for the present book. Steve Mason, *A History of the Jewish War: AD 66–74*, Key Conflicts of Classical Antiquity (Cambridge: Cambridge University Press, 2016) will likely be the standard book on the Jewish War for the foreseeable future.

1

Who Were the Soldiers in Palestine?

Thanks in large part to Jesus-movies and swords-and-sandals cinematic epics (e.g., *Ben-Hur*, *Masada*, *Spartacus*), there is a widespread perception that distinctively *Roman* soldiers infested Palestine during the life of Jesus—often signaled in such films by highbrow British accents in contrast with the unpretentious American dialect spoken by Jews. As deeply engrained as this image is in the popular consciousness, we will see that it is not entirely accurate. There were several different types of soldiers in the Roman East during the New Testament period and the differences between these soldiers were significant; the languages they spoke, the government they worked for, their relationship to the civilians they encountered, their pay, and many other specifics differed considerably. This book will work with a set of categories that aims to both clarify these differences and limit the scope to those present in the New Testament: legionaries, auxiliaries, royal soldiers, and the praetorian guard. There were other military men that will be noted on occasion who appear in the bible: the captains of the temple, bodyguards, etc. Conversely, soldiers who were not mentioned in the bible will not be discussed here: mercenaries, navy men, etc.

The aforementioned image of identifiably *Roman* soldiers occupying the land of Palestine operates on the assumption that the soldiers are legionaries. Legionaries differed from other soldiers of the early Roman period in several respects. First, legionaries were employed directly by Rome. Their allegiances were to the emperor and whichever general they served, not to any particular king, religious group, or province. Unlike most other soldiers, legionaries were Roman citizens before they were recruited. Though a legionary could theoretically come from any province within the empire, the requirement of Roman citizenship had consequences for demographics: legionaries were more likely to speak Latin than noncitizen soldiers, they were usually recruited from the most heavily Romanized cities and provinces, their citizenship held inherent prestige that afforded them privilege over both civilians and

1

other soldiers, etc. Legions primarily garrisoned in major imperial provinces, such as Syria, Pannonia, and postwar Judaea. With the exception of Egypt, all provinces with at least one legion were required to have a governor with senator status. Legions primarily consisted of infantry soldiers, with a few cavalry or archers present among their ranks. Roughly 30 legions were active at any given time within the empire and each consisted of approximately 5,400 soldiers and officers.

Roughly equal in number to the legionary soldiers across the empire were auxiliaries. Auxiliaries, like legionaries, served the government of Rome, but were divided into two distinct military types: cohorts and alae—infantry and cavalry, respectively—with a few mixed units termed *cohors equitatae* as well. Auxiliary soldiers were mostly noncitizens who were awarded Roman citizenship in exchange for military service. Consequently, auxiliary soldiers were significantly less Romanized than legionaries: auxiliary soldiers in the Roman East spoke the lingua franca of Greek and often local languages as well (e.g., Aramaic), typically with limited competence in Latin. The ethnic nature of these units led Rome to create many "specialist" cohorts (e.g., dromedary, archery, sling) that worked with combat methods familiar to one or another ethnic group. Though auxiliaries often served in major imperial provinces alongside legionaries, they also served in minor provinces as well. Thus, provinces and regions with a governor of Equestrian status (e.g., Raetia, Noricum, prewar Judaea) had no legions, but only auxiliaries. Until about 70 CE, many auxiliary soldiers were stationed in their home province; Judaeans were in Judaea, Syrians in Syria, etc. In addition to the Jewish War (66–73 CE), problems with soldiers' divided loyalties with the Revolt of the Batavi in Germania Inferior (69–70 CE) and the Year of the Four Emperors (68–69 CE) led emperors to actively undermine any remaining ethnic homogeneity in the *auxilia*, stationing soldiers outside their homeland in increasingly diverse units. Finally, auxiliaries were paid less than legionaries and did not receive all the bonuses granted to legionaries if they were successful in the same battle.

There were also royal forces that did not directly serve Rome, but were under the authority of a client king. The periphery of the Roman Empire was peppered with kingdoms allied with Rome that maintained their own militaries independent of the empire proper (e.g., Herod the Great's Judaea, Antipas' Galilee, Cleopatra's Egypt). These armies differed from kingdom to kingdom with respect to their hierarchies, pay scale, recruitment strategies, and so on. Rome occasionally expected kings to contribute soldiers to military campaigns as part of their reciprocal loyalty. Because kings could not offer their veterans Roman citizenship, the matter was irrelevant. With little invested in Romanness, royal soldiers spoke the local lingua franca and rarely had knowledge of Latin or other aspects of Roman culture.

Finally, the praetorian guard were a variety of soldier that, while fewer in number than these other types, are still relevant for New Testament studies. The praetorian guard were elite soldiers who served as the emperor's personal military force, being the only military unit within the region of Italia. Their proximity to the emperor was significant, as their loyalty might mean the difference between an attempted and a

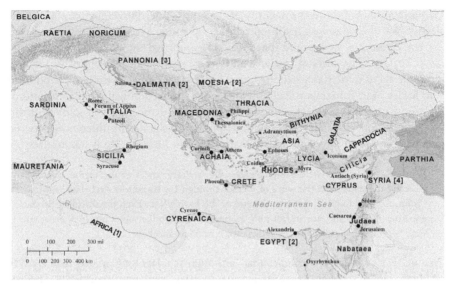

Figure 1.1. **Eastern Portion of the Roman Empire, circa 44 CE.** **Region names in all capital letters (e.g., SYRIA) were Roman provinces; combination of uppercase and lowercase letters (e.g., Judaea) indicates it was a client kingdom. Bracketed number indicates how many, if any, legions were garrisoned in that province as of 44 CE.**

successful assassination. The praetorian guard, insofar as it has limited importance for the study of the New Testament and differed significantly from other military units, will only be discussed as relevant.

Remembering the distinctions between these first three military units—legionaries, auxiliaries, and royal forces—is essential for discussion of prewar and postwar Palestine. The Jewish War (66–73 CE) was a catastrophic event for civilians in Palestine, regardless of their participation in the revolt against Rome. The destruction of the temple, the imposition of massive new military and administrative apparatus, widespread devastation, significant loss of life, among other factors, led to significantly different experiences of the military before and after the Jewish War. It is impossible to talk about the prewar and postwar life without attending to the details of these different units, especially auxiliaries and legionaries.

DEMOGRAPHIC MAKEUP OF PALESTINIAN SOLDIERS BEFORE THE JEWISH WAR

When Herod the Great reestablished the cities of Caesarea Maritima and Sebaste in 27 BCE, he granted his veterans land plots in these cities; Caesarea and the Sebaste quickly became the primary recruiting grounds for Herod's royal army.[1] Though both of these cities were in northern Judaea, their ethnic demographic was primarily Syrian, which came to be reflected in the military as well; though Jews and

Samaritans likely formed the majority of the army under Herod, by the time of the
Jewish War their numbers had been eclipsed by ethnic Syrians. Josephus notes that
Syrian Caesareans were proud that their kin comprised the bulk of the auxiliaries
(*A.J.* 20.176). The cities of Caesarea and Sebaste provided five cohorts of infantry
and one ala of cavalry as a standing army for Judaea. From the time of Herod until
the Jewish War, Palestinians had only incidental interaction with Roman legions or
its commanders. It is therefore useful to discuss the demography of soldiers of this
era as a distinct chronological period, despite some significant differences between
auxiliary and royal forces. Roman military historian Jonathan Roth summarizes:

> While there certainly were some changes, the military forces of the region remained
> basically the same from the reign of Herod, through his successors Archelaus, Antipas,
> Philip, Agrippa I and II, down to the end of the Jewish War. Even the so-called Roman
> garrison [i.e., *auxilia*] was in fact only a number of Herodian units put in Roman ser-
> vice. Most, perhaps all, of these soldiers were Aramaic speakers . . .[2]

Samuel Rocca likewise concludes that "although scholars long argued that Herod's
soldiers were for the most part foreign mercenaries, modern authorities . . . believe
that most of his troops were in fact Jews, and that Herod's army thus did not differ
much from the Hasmonaean army that preceded it."[3] This continuity was particularly
useful in ensuring stability through the political vicissitudes of the region. We will
see later that the Jewish War marked a significant change for military demographics,
shifting toward an army of occupation with an influx of foreign-born soldiers.

   Though Palestinian soldiers remained in their homeland after Archelaus was
banished and his principality annexed in 6 CE, some noteworthy changes occurred.
Since Judaea was now officially part of Rome, royal Herodian soldiers were subsumed into the Roman army as auxiliaries, though the Roman governors decided to
continue recruitment policies in place since the time of Herod the Great.[4] Military
diplomas—bronze tablets given to auxiliaries after completing their service as proof
of their citizenship—attest units named *cohors I Sebastenorum* and *ala Sebastenorum*,
an infantry and a cavalry unit named after the city of Sebaste.[5] However, the mere
fact that the cohort is given an ordinal number indicates that there was, at the very
least, also a unit named *cohors II Sebastenorum* and thus two cohorts and one ala.[6]
There is no reason to doubt Josephus' claim that the other three cohorts were recruited from Caesarea and Sebaste as well, presumably forming *cohortes III et IIII et
V Sebastenorum*.[7] Judaean governors continued recruiting troops from Caesarea and
Sebaste, who were then stationed within Judaea's borders. The military demographic
of the region shifted with the Jewish War, whereupon the Syrian legions became a
vital part of the social landscape, as will be discussed below.[8]

   In the Palestinian hinterlands, it was not practical to use Sebastene and Caesarean soldiers, so other locals were deployed to form military garrisons before the
war. Indeed, there was little reason for Judaea to supply soldiers to principalities
like Galilee and Batanaea. Herod transplanted some Idumaeans into Batanaea to
serve as a garrison against regional bandits (Josephus *A.J.* 16.285, 292) and also

fortified several villages to pacify Galilee's Hasmonaean sympathizers (Josephus *J.W.* 1.210). But as Herod's concern with Hasmonaean partisans declined so also did the strength of his forces in Galilee: a hoard of royal weapons at the Galilean city of Sepphoris fell out of use before Antipas began his reign (Josephus *J.W.* 2.56, 3.35–36). Josephus mentions other Herodian colonies at Hesbonitis, Gaba, and Idumaea, though little data survives regarding these sites. Herod set up a particularly important colony of Babylonian Jewish cavalry in Bathyra: its soldiers eventually served in the armies of Philip, Agrippa I, and came to comprise about half of Agrippa II's forces.[9] Three inscriptions attest a Jewish Bathyran cavalry commander named Diomedes under Agrippa II (§§30–32) and a different officer, Philip son of Jacimus, apparently mistreated shepherds of the Syrian desert (§145). The Bathyra colony included not only Jewish cavalry, but numerous Jewish civilians as well (*A.J.* 17.25–26; *syngeneis*). These civilians included women to marry soldiers and veterans to encourage the cavalry's proliferation. Caesarea and Sebaste were both major cities, so their military colonies served different functions from those of Bathyra, but the cities nevertheless encouraged veteran settlement and marriage. Even though Caesarea and Sebaste were primarily Gentile, we will see that Caesarean Jews also served in the Roman army (§199, §294, §296).

While nearly all the forces of prewar Palestine were native, there is some evidence of foreigners in their ranks as well. Josephus mentions Thracian, Galatian, and German soldiers attending Herod's funeral (*A.J.* 17.198, *J.W.* 1.672), but there is only limited evidence of their use during his reign and none among his successors. Rather, we find that Herod's army is treated as Judaean in Josephus' works.[10] The heavy enrollment of Palestinian soldiers is also evident in the epigraphic record for both Herodian and Roman forces.[11] 1) The aforementioned Diomedes son of Chares was in the Babylonian Jewish cavalry at Bathyra, serving as eparch under Agrippa II and in *cohors Augusta* during the Jewish War (§§30–32). Despite his Judaism, Diomedes openly supported Batanaea's local (i.e., non-Jewish) religions.[12] Josephus names a few other members of this colony, though Philip son of Jacimus is the only other one attested epigraphically (§145). 2) Herod son of Aumos was an Idumaean colonist in Batanaea who served as a high-ranking officer under Agrippa II (§23). 3) Lucius Cornelius Simon was a Caesarean Jew that served Rome in the Jewish War (§294). 4) Publius Aelius Mercator was a Caesarean local who fought in *legio I Adiutrix* during the Bar Kokhba War (§199). 5) Aelius Silvanus was a Jerusalem native who fought in *legio II Adiutrix* during the Bar Kokhba War as a centurion (§200). 6) Bar-Simsus Callisthenes was a Jewish man from Caesarea in *cohors I Vindelicorum miliaria* (§296). 7) A soldier named Matthew garrisoned in Herodium, though it is not clear whether he was a royal soldier or Roman auxiliary (§120). 8) Titus Flavius Iuncus was an officer in various auxiliary cohorts and *legio X Fretensis*, born in Neapolis (§§192–193). 9) One inscription attests two soldiers from Caesarea and other soldiers from Gaza and Anthedon in *legio III Augusta* during the early second century (§184). 10) Ausos son of Aios was an Arabian Batanaean who joined Agrippa II's army (§34). 11) Matthias son of Polaius, a Jewish man from somewhere in Syria-Palestine, served in *legio I Adiutrix* (§293). There is also extensive evidence of soldiers recruited from the Decapolis.[13]

Research concerning Jews in the army is hindered by historical accidents that prevent us from knowing much about most soldiers. Aside from obvious problems of fragmentary inscriptions and the unpredictability of archaeological finds, the question of soldiers' names impedes identification of origin and ethnicity. While many biblical scholars assume that soldiers with Roman names must have been Roman citizens, evidence suggests otherwise: one papyrus written in 103 CE indicates that some auxiliaries received Romanized names (i.e., *tria nomina*) shortly after recruitment, even before training completed.[14] Because some soldiers changed their name shortly after recruitment, the mere act of joining the military often obscured soldiers' ethnic and geographic origins. Benjamin Isaac thus observes a few obvious instances where soldiers from the Decapolis dropped their Semitic birth name to take up a Roman one.[15] Consequently, there is no way to know the birth name of, say, the auxiliary soldier Domitius son of Domitius (who was from the Decapolis city of Philadelphia), since he followed the convention of adopting the name of his commander (§295, in this case the Syrian legate Domitius Corbulo).

Thus, while Rome did not conscript Jews into military service against their will, there is no indication that this prevented them from serving on their own accord. In addition to the Palestinian Jews mentioned above, Jewish soldiers born elsewhere in the empire also warrant attention. 1) Best documented is a collection of letters from an Egyptian soldier named Julius Apollinarius to his family (§§167–171). Julius served in *legio III Cyrenaica* and had a Jewish grandmother named Sambathion.[16] Julius' father was also a member of that legion. 2) Several tax receipts of a Jewish decurion named Jesus around the turn of the first century survive from Edfu in Egypt (§§172–179). 3) Another tax receipt from Egypt attests a Jewish centurion named Hananiah in the early second century (§183). 4) A somewhat later inscription may honor the descendent of another Egyptian centurion named Benjamin (§71). 5) A diploma was issued to Aggaeus Bar-Callippus, a Jewish veteran who retired to the Syrian city of Samosata (§257). 6) We should not forget the famous example of Tiberius Julius Alexander, governor of Judaea and Egypt, a Jewish officer who led the assault on the Jerusalem temple in the Jewish War (§§364–373).

But how did the Jewish religion fit into the Roman army? Though many commentators assert or assume that soldiers were obligated to worship the emperor or the Roman pantheon, the Roman military generally respected troops' religious practices.[17] There is little reason to suppose that Jewish soldiers no longer identified as Jews, even if it became complicated at times: a Jewish soldier named Matthew tended to the pigs at Herodium (§120); even if Matthew's labor rendered him unclean by the prevailing ritual standards of Judaism, there is no reason to infer that he no longer cared about Jewishness. Several veterans kept some part of their Semitic birth names even when eligible for the *tria nomina*; military diplomas attest auxiliary veterans with the names of Bar-Callippus (§257), Matthias (§293), Lucius Cornelius Simon (§294), and Bar-Simsus (§296), each of whom could have abandoned it in favor of a more Romanized name. Ethnic heritage thus served as a continuing point of identity among Jews and other Palestinians in the military.

It is thus important to avoid involvement in the authenticity politics concerning first-century debates as to whether Jews in the military were "really" Jewish or not. Andrew Schoenfeld is particularly critical of the assumption that Jews in the Roman army were apostates and thus effectively Gentile.[18] Jewish practices varied considerably, such that one person's piety might be another's heresy. To be sure, upon reintegration in civilian life, certain military practices may have compromised their adherence to Judaism in the eyes of some (e.g., Sabbath labor, table fellowship). However, in-group debates among early Roman Jews about the (in)sincerity of Jews in the military and the parameters of "authentic Judaism" should not be mistaken for scholarly definitions of these terms. We cannot ascertain the degree to which Jewishness or Judaism as a religion remained important to these men upon joining military service and it would be disingenuous to pretend otherwise.

THE BENEFITS OF SERVICE FOR PALESTINIAN RECRUITS

Since military service was not mandatory for anyone in Palestine, one may wonder what factors led men to volunteer. But in addressing this issue, one should avoid the inference that military service resulted from the self-selection that characterizes job-choice today: the vast majority of Palestinian recruits were born and raised in military colonies such as Bathyra and Gaba, or heavily militarized cities like Caesarea and Sebaste. As young men, they were socialized into military values, customs, and discourses—their fathers, uncles, brothers, or neighbors may have been soldiers, rendering the soldiering life a familiar and appealing career. The Roman writer Seneca cites an example that is valuable for the expectations it reveals about enrollment of able-bodied youth.

> Fabianus says, as my parents also saw, that as a boy at Rome he had the stature of a large man. But he swiftly departed this life and no one of discernment could have failed to say that he would soon die; for he was not able to reach that age of life which he had attained prematurely. . . . Although his stature, his fairness of form, and his particularly robust physical constitution marked him out as one born for service in the field as a soldier, he refused to follow the colors so that he might remain with you. Consider, Marcia, how seldom women who live in a different dwelling from you see their children. Reflect upon the fact that all those years count as lost years and years passed with never ending anxiety for mothers who have sons in the army. (*Marc.* 23–24; trans. Fred Baxter)

Seneca, writing a letter of consolation to the deceased youth's mother, expresses pity for the bereaved; Seneca had earlier voiced anxiety over his own son's nearly assured military service (*Marc.* 9). We might think of military recruitment less as a matter of mandatory inevitability than as a process in which employment strategies were inseparable from the other strategies of social reproduction: many young men were socialized into norms that rendered military service appealing, not only because of military service's benefits, but because their identity and values had already been produced within the frame of military masculinity.

What aspects of this frame might appeal to men of early Roman Palestine? The Roman army underwent "Marian Reforms" in 107 BCE, including a provision that opened military service to men regardless of wealth; among other requirements, a man previously needed to own property totaling at least 3,500 sestertii and needed to supply his own weapons. The Marian Reforms led to a considerably larger and economically destitute military: most of the soldiers in legions, *auxilia*, and royal armies were recruited from the lower classes by the first century CE.[19] Palestine was no exception. Josephus reports "though [the Syrian Caesareans] were inferior regarding wealth [compared to Jewish Caesareans], they prided themselves in the fact that they were most of the Romans garrisoned there, whether Caesarea or Sebaste. . . ."[20] More revealing is the humble pay that soldiers received in exchange for their labor.[21] Pay amounted to 750 sestertii per year for low-ranking auxiliaries before a raise in 84 CE. About half of the pay went to living expenses, such as food and clothing; thus, while soldiers nominally earned about two sestertii per day, after deductions it was closer to one sestertius. This pay compares unfavorably with the wages of manual laborers in the first and early second centuries CE. Sources are limited for Palestine, but reports indicate four sestertii per day was typical: Rabbi Hillel worked for four to eight sestertii per day (b. Yoma 35b; 'Abot R. Nat. 27b), Matt 20:1–15 and the Talmud (b. 'Abod Zar. 62a) both report four sestertii per day as common. By way of comparison, *P.Lond.* 131 reports the equivalent of one-and-a-half to two sestertii per day for laborers in Egypt in the first century CE and Cicero (*Rosc. com.* 28) reports twelve per day in Rome in 80 BCE.[22] Auxiliaries were not the only soldiers paid poorly. For a year's labor with extensive combat, Herod paid his men 450 sestertii.[23] Josephus notes that Herod's Nabataean rivals attempted to sow sedition among his soldiers by capitalizing on resentment over wages (*A.J.* 15.353). Even legionaries stationed in Judaea after the Jewish War spent nearly their entire pay on necessary expenses, ending up penniless after payday (§22). Tacitus' assessment of legionary life in the early years of Tiberius' reign is relevant: "In fact the whole trade of war was comfortless and profitless. Ten asses [i.e., 2.5 sestertii] a day was the assessment of body and soul: with which they had to buy clothes, weapons and tents, bribe the bullying centurion, and purchase a respite from the duty." (*Ann.* 1.17). Given that soldier's wages were meager—especially after mandatory deductions—most men could not expect to find wealth in the military. Palestine was not subject to the variety of hostile combat that predicated looting (unlike other frontier territories) until the Jewish War, precluding booty as a source of income. Though there were land grants for both Herod's veterans and the postwar garrison, it is unlikely that these farmsteads were sufficient for anything beyond subsistence.[24] Auxiliaries and royal forces were also ineligible for incidental bonuses, like *donativia* for exceptional combat and *praemia* for retirement (Cassius Dio 73.8.4). Except for commissioned officers, there is no reason to believe that eventual wealth lured anyone into the soldiering life.

What benefits, then, did the military proffer soldiers, if not financial prosperity? In answering this question, one must remember that impoverishment is not simply a matter of finances, as construed in modern conceptions of an income-determined "poverty threshold." Roman historians Peter Garnsey and Greg Woolf instead encourage us to think of the poor as "those living at or near subsistence level, whose prime concern it is to obtain the minimum food, shelter, and clothing necessary to sustain life, whose lives are dominated by the struggle for physical survival."[25] Garnsey and Woolf argue that income was one among many indices of poverty in antiquity, including underemployment, housing insecurity, health care, premature mortality, nourishment, and legal status. Some recent estimates place 90 percent to 99 percent of the Roman Empire's population within Garnsey's understanding of impoverishment (with roughly the same percentage in Palestine in particular).[26] The military offered a means of stability and even modest surplus in light of the prevailing—even endemic—insecurity of life's necessities for most of the Roman world, thus explaining why potential recruits were typically found among the destitute.

To start, military service offered twenty-five years of guaranteed employment in the *auxilia* and comparably long-term employment in royal armies. The Roman army professionalized into a standing force after the Second Punic War at the turn of the second century BCE, with the legions no longer comprising temporary volunteers, but men for whom it was a career. Rome formed its *auxilia* in 23 BCE and Herodian client kings consistently kept a standing army as well. Soldiers did not need to worry about long-term employment as did day-laborers, crop failure as did farmers, or the availability of contracts as did tradespersons. Moreover, income and employment were assured year-round from an employer with reliable income, as opposed to the estimated 225 working days a year for day-laborers of antiquity.[27] There were further financial benefits of military service: housing, whether in a fortress or some other barracks, was reliable and free. Consequently, it was rare for soldiers to incur the degree of debt that was common among farmers and other laborers. While soldiers had to pay for many of their expenses (e.g., food, clothing), guaranteed shelter and near-complete freedom from the threat of debt were two considerable benefits.

Soldiers were also assured nutritious food. Archaeologists have discovered, for instance, numerous "bread stamps" throughout Palestine (§§83–91). Each centurion designated one soldier under him as the century's baker. This man was responsible for producing bread for his unit and then stamping it to indicate the unit for which it was produced. Soldiers throughout the empire consumed excellent food in comparison with their civilian counterparts. A number of military shopping lists discovered at British Vindolanda indicate the garrison's high-quality food: "two modii of bruised beans, twenty chickens, a hundred apples, if you can find nice ones, a hundred eggs or two hundred if they are for sale there at a fair price. [. . .] eight sextarii of fish-sauce [. . .] one modius of olives [. . .]." (*T.Vindol.* 302) Diversity of food is also evident in fortress excavations, which attest many numerous meat products, including oxen, sheep, pigs, goats, deer, oysters, and various fowl.[28] Josephus likewise describes Herod the Great as providing his soldiers "grain, wine, oil, cattle, and all other provisions. . . ." (*A.J.* 14.408; cf. *J.W.* 1.299). Many papyri, however, indicate that soldiers contacted family for additional food, though it is often unclear why they did so; in some instances, requests for food merely results from squandered pay or a diet that was repetitive and bland.[29] It is also worth noting that a uniform diet was not imposed upon all soldiers: the eating habits of ethnic auxiliaries and royal soldiers were accommodated if possible.[30] However, it was apparently easier for some soldiers to adjust their own eating habits to what the army offered: a Jewish soldier at Herodium was a swineherd for his unit (§120).

While we often imagine the soldiering life was life-threatening, this was rarely so in prewar Palestine. There were indeed insurrections and tensions on occasion, but they did not require the attention of the entire military. There were few major combat operations, soldiers rarely left their homeland, regular exercise aided longevity, and medical care was provided. Rome took the health of its soldiers seriously. Soldiers were permitted sick days for rest and recuperation (see, e.g., *T.Vindol.* 154) and each auxiliary cohort and ala had its own medical staff.[31] Doctors also granted discharges to ill or injured soldiers. If a soldier were permanently injured and incapable of service, they were nevertheless given full benefits as though they had completed their full term (*AE* 1932.27). Medical care for soldiers in Judaea was no exception.[32]

Finally, there is the complex issue of Roman citizenship, which had many benefits. In 212 CE, the emperor Caracalla granted citizenship to all free men in the Roman Empire. But before this, Roman citizenship was mostly limited to men who inherited it from their fathers. Since auxiliaries were noncitizens, they were awarded citizenship after a requisite term of military service, proof of which was given to them in the form of a military diploma (see Figure 1.2). Citizenship was initially granted after thirty years of service, but was reduced to twenty-five years and expanded to include auxiliaries' wives and children around 50 CE.[33] This was a considerable boon and reason enough to join the *auxilia*. First, auxiliary soldiers' marriages were recognized by the state, unlike legionaries who were banned from marriage while in active duty.[34] The names of wives and children were included in auxiliary diplomas until 140 CE; nearly half of the diplomas indicate soldiers took advantage of this opportunity.[35] Thus, while legionaries were unable to legally marry during their prime years (though they are well attested as clients of sex workers), auxiliaries were permitted to have their own families, with legal rights conferred upon them as well. Second, Roman citizens held enormous privilege, being guaranteed trials, the right to sue, assurances against beatings, freedom from certain taxes, among many other benefits. Auxiliaries could expect such benefits, but client kings lacked the authority to grant citizenship or legal marriage. For this reason, auxiliary service was inherently more desirable than service under a client king. Roman citizenship was rare even among the upper classes of the Roman East,

Figure 1.2. Roman Military Diploma. Portion of a military diploma issued in 90 CE to a soldier who last served in the province of Judaea (§208). The soldier's name, rank, and unit has been lost.

Photo credit: Or Fialkov, used by permission.

which made it all the more valuable for soldiers, aiding social advancement in a way that soldiers might not expect where citizenship was more common.

The factors that might lead a typical Palestinian man to join the army should now be clear. In the words of Sara Phang, "from an absolute perspective, compared to the masses, soldiers were privileged, though less so than the elite groups."[36] Relative to soldiers' civilian peers, military service offered an escape from the precarity that defined daily life. Food security, healthcare, housing, freedom from substantial debt, and legal assurances could be taken for granted in military life—things that can only be construed as "privileges" insofar as they were not available to all, as they are basic characteristics of a tolerable life. Thus, if one is to characterize recruitment in military colonies and militarized cities as "voluntary," it would only be in the sense that soldiers were rarely conscripted against their will. This was not a matter a potential recruit debated at length with his family; if a recruitment officer expressed interest in enlisting a young man, he was rarely in a position to decline the offer. In that youth living in military colonies were habituated into military norms and found a seemingly natural "fit" for their tastes, practices, and ambitions in soldiering, it was hardly a decision brought about by their own free accord; rather, the cultivation of desires and expectations from their social context played a significant role in enlistment. If one hoped to benefit from the rigged game of ancient economics, military employment provided one of the safest bets.

LEGIONARIES SERVING IN JUDAEA BETWEEN
THE JEWISH WAR AND THE BAR KOKHBA WAR

The canonical Gospels and Acts were all written after the Jewish War (66–73 CE). Though the trauma of the Jewish War is well known, it had a significant impact on the demographics of the military in Palestine. The lenient policy permitting a soldier to garrison in his homeland was abolished after the Jewish War by the Flavian emperors, whose generals stationed auxiliary regiments away from native soil, expended little effort to maintain ethnically homogeneous cohorts, and more actively encouraged soldiers to assimilate to Roman cultural mores. Consequently, Roman auxiliaries recruited from Judaea remained in Palestine until the Jewish War, after which they were dispersed throughout the empire (Josephus *A.J.* 19.366). The postwar auxiliary units of Judaea varied, including units from Thrace, Thebes, Britain, Iberia, among several other far-off provinces.[37] As one might expect, their origination outside Syria and Palestine contributed to the perception that they were foreign by Palestinian denizens.

It is also at this point that legions became part of the Palestinian landscape.[38] As noted earlier, Roman citizenship was a prerequisite for legionary recruitment. Though it is common to suppose that most citizens were Italian or aristocrats, this was not the case. *Legio X Fretensis* was the only legion garrisoned in Palestine from the end of the Jewish War until late in Trajan's reign, when it was supplemented with *II Traiana* and eventually *VI Ferrata*. Before the Jewish War, *Legio X Fretensis* garrisoned in Syria, with the exception of combat missions to Judaea after its annexation in 6 CE and the Parthian campaigns 58–63 CE. Though few of its legionaries were Palestinian by birth, most of the men in *legio X Fretensis* were nevertheless recruited from the Roman East, where the legion had long been located. Soldiers of *legio X Fretensis* were primarily drawn from Syria and were only partially Romanized (e.g., spoke Greek as a native language with some Latin familiarity, modest use of Roman social conventions). The Syrian legions, while certainly not alien in their demographic, nevertheless lent a newfound foreignness to the military in Palestine immediately after the Jewish War; compared to the local policing of the prewar period, the postwar garrison was readily understood as an occupying force. Even less familiar were the postwar auxiliaries coming from even further away (§160, §§202–209). As a province where Latin and Roman citizenship were extremely rare before the War, Judaean civilians experienced legions as a *Roman* intrusion. Thus, one finds an increase in Latinisms in Palestinian texts composed in Greek and Aramaic, usually words with specifically military meanings; merchants and craftspersons near garrisons likewise catered more directly to Roman consumption habits than they had to the *cohortes Sebastenorum*.[39]

Legionaries received slightly better pay than auxiliaries, but most of what was said about auxiliaries could be reiterated here, with a few noteworthy exceptions. First, legionaries were somewhat less likely to be impoverished, though before recruitment they seem to have reflected general demographics of their society. Second, because eventual citizenship could not act as a lure for recruitment, legions were awarded

Figure 1.3. Plundering the Jerusalem Temple. Replica of detail in the Arch of Titus (83 CE), depicting the Roman legionaries plundering the Jerusalem temple following its destruction in 70 CE.
Photo credit: Shem Tov Sasson, used by permission.

donativia of varying sums and legionaries all received *praemia* of 3,000 denars after retirement. In the tumultuous Roman Civil War of 69 CE, each of the four emperors promised donatives to the legions to secure their allegiance; financial assurances were thus much greater for legionaries than auxiliaries. Third, plunder was a far more common method of income among legionaries than auxiliaries, especially those auxiliaries stationed in their home province. Looting was commonplace during warfare and Josephus records a number of incidents (e.g., *J.W.* 2.494, 503–506, 508–509, 646) including the plundering of the temple itself (*J.W.* 5.550–566). Finally, legionaries were more fluent in Latin, though Greek continued to function as the lingua franca among civilians in the Roman East. This inevitably led to occasional barriers in communication, though the prevalence of Greek inscriptions and papyri from legionaries indicates that they sometimes preferred that language over Latin.[40]

RECRUITMENT AND PROMOTION
OF CENTURIONS IN PALESTINE

It is worthwhile to contrast the life of common soldiers with that of noncommissioned officers,[41] such as centurions—a rank that features prominently in the New Testament. Centuries were the basic subunit in auxiliary and legionary forces (see §§83–91; cf. §26, §60), with a centurion at the head of each. Herodian royal armies organized along Roman lines and so centurions are also attested among their ranks.[42] These units were nominally 100 men strong, but were around 80 in practice.

Since Rome expected low-level officers to communicate directly with their troops, it was common for auxiliaries to remain in or near their homeland. The various dialects of Western Aramaic were mutually intelligible, but centurions, being the highest ranking of noncommissioned officers, needed to both understand their Latin- or Greek-speaking superiors and direct their Greek- and Aramaic-speaking subordinates (see, e.g., Josephus J. W. 4.37–38). This was less necessary after the Jewish War, when the relatively friendly interactions of the prewar period were exchanged for policies typifying an army of occupation.

We saw above that Jews and other Palestinians served as officers, including centurions, in both the *auxilia* and legions.[43] Josephus only provides the name of one auxiliary centurion garrisoned in Judaea: Capito (*J. W.* 2.298–300), a Latin cognomen meaning "big head."[44] Though some might argue this is evidence of Romans in the Judaean *auxilia*, Steve Mason is hesitant to make too much of this name, as Mason sees a trope in Josephus' *Jewish War* of emphasizing the Roman element among auxiliary officers.[45] Josephus mentions a centurion named Arius, but it is not clear whether he was auxiliary or legionary, even if his name is Greek.[46] Epigraphic data does not provide much additional help regarding centurions from Palestine, though Lucius Obulnius (§§12–18) and Aelius Silvanus (§200) both have inscriptions erected in their name. Little is known about the former and the latter was a legionary. That said, Archieus (§38) was a centurion under Agrippa II until his death, whereupon Archieus transferred into the Roman army before retiring as a *strategos* under Trajan. Many other centurions in Palestine during and after the Jewish War are also known (e.g., §4, §21, §28, §§30–32, §35). Somewhat frustratingly, historical accident has left us with inscriptions primarily from postwar legionaries and only rarely the Palestinian auxiliaries or prewar client kings' forces.

In addition to these noncommissioned officers, there were commissioned officers who were sufficiently reputable that they served in both Roman and Herodian armies. Titus Mucius Clemens had the rank of eparch in both royal and Roman forces, a pattern that also characterizes Diomedes' transfer between Agrippa II's army and *cohors I Canathenorum Augusta* and Archieus' incorporation into Trajan's imperial army.[47] It is likely that Agrippa II's support of the Roman campaigns in both Parthia and Judaea resulted in the temporary incorporation of his men into auxiliary cohorts and thereby necessitated ad hoc conformity to Rome's military hierarchies.[48]

Tiberius Julius Alexander, an Alexandrian Jew and nephew of Philo, served as governor of Judaea 46–48 CE and partook in the siege of Jerusalem as Titus' chief of staff.[49] Tiberius had enviable social connections that aided his personal and political ambitions, including a father who was a prominent politician and a brother who married Agrippa I's daughter (and thus Agrippa II's sister) Berenice. Berenice had an affair with the eventual emperor Titus; so it is likely not mere coincidence that Tiberius Julius Alexander joined Titus in leading the assault on the Jerusalem temple in the Jewish War. Whatever the proceedings of the typical equestrian career, his familial and social connections likely facilitated his rise to particular prominence. Family associates appear to have fostered the career of another diaspora Jew named

Antiochus, a ruthless officer who commanded a detachment of Roman soldiers shortly before the Jewish War.[50] Antiochus also came from an influential family, as his father was a magistrate in Antioch. Finally, it is worth noting that Diomedes the centurion was part of a prominent family (§§30–32; Josephus, *Life* 35, 37), being related to Philip the governor of Gamala (cf. §145). Though there was a legal distinction between imperial and military officials, there was substantial overlap in practice—with provincial governors acting as de facto military commanders and military officers taking on administrative roles as need be. Thus, while Tiberius Julius Alexander and others held prominent political positions, his experience in gubernatorial offices cultivated familiarity with military practices as well; this, one should not be surprised, was by design.

How might a soldier ascend the ranks to become a centurion?[51] Provincial governors were ultimately responsible for promoting men to the rank of centurion. Other officers played an integral role in the selection process by recommending soldiers in whom they saw promise. This is evinced in the case of Apollinarius, a legionary with a Jewish background, who requested a promotion to the governor's personal secretary; though no positions were open, the governor made him secretary of the legion with the possibility of further advancement (§167). The most typical method of advancement was to pass slowly through the ranks, from *pedes* to *immunis* to *tesserarius* to *optio* to *centurio*.[52] Fundamental to the promotion process was performance of exceptional deeds and support of one's troops. One soldier in Egypt said that he was "promoted [to centurion] because of brave deeds and the detachment's support. . . ."[53] Epistolary evidence from Egypt suggests that literacy was taken into consideration during the promotion process. In one letter, a recent recruit named Apion thanked his father because "you educated me well and as a result of that I hope to be advanced quickly" and a different Egyptian soldier explicitly attributed his own advancement to education (*BGU* 423; cf. §168). That said, "quick" promotion was entirely relative, as a survey of military careers indicates it typically required ten to twenty years of exceptional service to reach the centurionate. In rare circumstances, most notably during civil wars, units would elect their own centurions, which Eric Birley suggests encouraged loyalty to both the unit and its commanders.[54]

Many centurions were appointed at that rank if they came from relatively high social strata, but most were promoted on the basis of individual characteristics. Personality, linguistic facility, socioeconomic origins, interactions with superiors and subordinates, as well as other individual factors were central to the process. We will see in the next chapter that this difference played a pivotal role in the divergent ways in which civilians interacted with common soldiers as opposed to centurions.

CONCLUSION

Before the Jewish War (66–73 CE), the vast majority of soldiers in Palestine were local recruits. Parts of Palestine were client kingdoms and other portions were a Roman province: the former employed royal soldiers and the latter auxiliaries, as was done throughout the empire. Prewar soldiers were largely indistinguishable from their peers when out of uniform: ethnically, linguistically, economically, religiously, etc., they were unremarkable denizens of Palestine. This changed with the Jewish War, which not only destroyed much of Palestine, but was a watershed moment in military demographics: Judaean auxiliaries were transferred to other provinces, an influx of foreign auxiliaries replaced them, and legionaries were introduced to the region. Thus, while before the Jewish War the military comprised a local policing force, after the war it became an army of occupation with soldiers that—by comparison with the prewar garrison—seemed quite Roman. Both before the war and after, soldiers were often poor and military employment helped mitigate the precarity of life.

NOTES

1. See, e.g., Josephus *A.J.* 15.296, 20.122, 20.176, *J.W.* 1.403, 2.52, 2.58, 2.74, 2.236, 3.66. This is a controverted topic. Steve Mason, Jonathan Roth, and others contend that the Judaean *auxilia* was primarily Jewish or Samaritan. Against this perspective, it should be observed that Josephus consistently claims that the most prominent ethnic group in the Judean military were Syrians recruited from Caesarea and Sebaste. Though this is integral to Josephus' narrative of *stasis* via ethnic conflict and radicalized parties, there is little reason to suppose ethnic Syrians did not form a majority or substantial minority of Judaean auxiliaries.

2. Jonathan P. Roth, "Jewish Military Forces in the Roman Service," in *Essential Essays for the Study of the Military in New Testament Palestine*, ed. Christopher B. Zeichmann (Eugene: Pickwick, forthcoming).

3. Rocca, *Army of Herod*, 13; cf. Shatzman, *Armies*, 170–216. Contrast Aryeh Kasher, *Jews and Hellenistic Cities in Eretz Israel: Relations of the Jews in Eretz-Israel with the Hellenistic Cities during the Second Temple Period (332 BCE–70 CE)*, TSAJ 21 (Tübingen: Mohr Siebeck, 1990), 209–214.

4. See the discussion of this process in Ian Haynes, *Blood of the Provinces: The Roman Auxilia and the Making of Provincial Society from Augustus to the Severans* (Oxford: Oxford University Press, 2013), 116–119. Evidence concerning the effects of annexation on royal armies is scarce, though Judaea appears to have been very simple in its annexation of military forces. After Nabataea was annexed in 106 CE, its soldiers were divided up into *cohortes I–VI Ulpiae Petraeorum*, a transition aided by the Nabataean kings' modelling of theirs upon Roman military hierarchies (D. F. Graf, "The Nabataean Army and the *Cohortes Ulpiae Petraeorum*," in *The Roman and Byzantine Army in the East*, ed. Edward Dąbrowa (Krakow: Drukarnia Uniwersytetu Jagiellońskiego, 1994), 265–305). The annexation of Pontus and the Gallic Julii (John Drinkwater, "The Rise and Fall of the Gallic Iulii: Aspects of the Development of the Aristocracy of the Three Gauls under the Early Roman Empire," *Latomus* 37 (1978): 824–831) involved direct transfer of soldiers from the existing army to the Roman military, like Judaea (Tacitus, *Hist.* 3.47).

5. The standard argument on the name of the auxiliary units in Judaea remains Emil Schürer, *The History of the Jewish People in the Age of Jesus Christ (175 B.C.–A.D. 135)*, trans. Geza Vermes, Fergus Millar, and Matthew Black, Revised English ed., 3 vols. (Edinburgh: T&T Clark, 1973–87), 1.363–365. See also the texts cited in note 1 above.

6. §§210–227, §§229–231 on *cohors I Sebastenorum*. §§232–245 on *ala Sebastenorum*. It is likely the other cohorts were devastated in the Jewish War (see, e.g., Josephus *J. W.* 2.430–437).

7. *cohors Ascalonitanorum* is also attested; this is not surprising, given that Ascalon was a free city operating independently of Judaea and so had a discrete garrison (§136, §161; cf. §§225–227, §229, §242, §§246–259). Josephus mentions a garrison of one cohort and one ala in Ascalon on the eve of the Jewish War (*J. W.* 3.12), though no further evidence of *ala Ascalonitanorum* survives.

8. After the Jewish War, the soldiers of *cohors Sebastenorum* and *ala Sebastenorum* were no longer drawn from Judaea; for instance, §221 (160 CE) was issued to a Cilician soldier.

9. Josephus *A.J.* 17.23–31; cf. Tg. Ps.-J. Num 34:15. See the fuller studies in Shimon Applebaum, *Judaea in Hellenistic and Roman Times: Historical and Archaeological Essays*, SJLA 40 (Leuven: Brill Academic, 1989), 47–65; Getzel M. Cohen, "The Hellenistic Military Colony: A Herodian Example," *Transactions of the American Philological Association* 103 (1972): 83–95; Christopher B. Zeichmann, "Herodian Kings and Their Soldiers in the Acts of the Apostles: A Response to Craig Keener," *Journal of Greco-Roman Christianity and Judaism* 11 (2015): 178–190.

10. Shatzman, *Armies*, 186 has a thorough treatment of Josephus' testimony. He cites the Nabataean-Judaean War of 31–30 BCE (*A.J.* 15.111–146, *J. W.* 1.366–384) and the rebellion upon Herod's death in 4 BCE presumes a great number of Jews in his army (*J. W.* 2.52).

11. For further discussion, see Jonathan P. Roth, "Jews and the Roman Army: Perceptions and Realities," in *The Impact of the Roman Army (200 BC–AD 476): Economic, Social, Political, Religious, and Cultural Aspects*, ed. Lukas de Blois and Elio Lo Cascio, Impact of Empire 6 (Leuven: Brill, 2006), 409–420; Andrew J. Schoenfeld, "Sons of Israel in Caesar's Service: Jewish Soldiers in the Roman Military," *Shofar* 24/3 (2006): 115–126; Samuel Rocca, "Josephus, Suetonius, and Tacitus on the Military Service of the Jews of Rome: Discrimination or Norm?" *Italia* 20 (2010): 7–30, but especially insightful is Raúl González Salinero, "El servicio militar de los judíos en el ejército romano," *Aquila Legionis* 4 (2003): 45–91.

12. While worship of a single deity was regulated by Jewish social norms, it was not always observed strictly, as Jews commonly partook in cultic rituals with other gods. See Paula Fredriksen, "Review of N.T. Wright, *Paul and the Faithfulness of God*," *CBQ* 77 (2015): 390.

13. See, e.g., §33, §147, §158, §162, §198, §204, §209, §294, §295, §363. See also §184 with one soldier from Scythopolis and two from each Damascus and Hippos.

14. *P.Oxy.* 1022. See the discussions of the text in Roy W. Davies, *Service in the Roman Army* (Edinburgh: Edinburgh University Press, 1989), 17–18 and *RMR* 87. Cf. *BGU* 423 and Nigel Pollard, *Soldiers, Cities, and Civilians in Roman Syria* (Ann Arbor: University of Michigan Press, 2000), 128 on *RMR* 1. Auxiliaries in *cohors I Hispanorum* also had *tria nomina* in 83 CE (*IGR* 1.1337).

15. Benjamin Isaac, "The Decapolis in Syria: A Neglected Inscription," *ZPE* 44 (1981): 72–73 cites §147 and §295. Recall also the names of Agrippa I and II: Gaius Julius Agrippa and Marcus Julius Agrippa, neither of which suggests Jewishness.

16. Julius' references to "the gods" suggest that this Judaism was peripheral to his identity practices (§§167–168); this may also be the case with Diomedes who, though a Zamarid, dedicated a statue to Zeus Beelbaaros (§32). Note also that while Julius' father was a legionary, it is not clear whether Sambathion was Julius' maternal or paternal grandmother.

17. Allan S. Hoey, "Official Policy towards Oriental Cults in the Roman Army," *Transactions of the American Philological Association* 70 (1939): 456–481; Oliver Stoll, "The Religions of the Armies," in *A Companion to the Roman Army*, ed. Paul Erdkamp, BCAW (London:

Blackwell, 2007), 464–471; cf. the *locus classicus* Tacitus *Germ.* 43. A revealing example is the archive of Claudius Terentianus (enlisted in the *classis Alexandriae* 110–136 CE; *P.Mich.* 8.476–481). In correspondence between Claudius and his friends, religious vocabulary is regularly invoked, but Roman deities are never named; Claudius and his friends instead opt for the god Serapis and associated deities. Likewise, the soldiers of *legio I Minervia* located in Germania Inferior had no interest in Roman deities, instead worshipping local goddesses such as the Matronae Aufaniae (Rudolf Haensch, "Inschriften und Bevölkerungsgeschichte Niedergermaniens: Zu den Soldaten der *legiones I Minervia* und *XXX Ulpia Victrix*," *Kölner Jahrbuch* 33 (2001): 89–134).

18. Schoenfeld, "Sons of Israel," 116–117.

19. Tacitus' unhappy description of recruits under the emperor Tiberius is typical: "Volunteers were not forthcoming, and even if they were sufficiently numerous, they had not the same bravery and discipline, as it is chiefly the needy and the homeless who adopt by their own choice a soldier's life" (*Ann.* 4.4).

20. Josephus *A.J.* 20.176. Inscriptions indicate that names common among the lower classes were prevalent among soldiers serving in Judaea; see, for instance, Leah Di Segni and Shlomit Weksler-Bdolah, "Three Military Bread Stamps from the Western Wall Plaza Excavations," *Atiqot* 70 (2012): 21–31 on §§85–87; but contrast the exceptional instance of §67—though the soldier's rank is unknown.

21. The question of pay for auxiliaries was long contested, but Michael P. Speidel's suggestion that it was five-sixths of a legionary's pay ("The Pay of the Auxilia," *JRS* 63 (1973): 141–147) has been confirmed by his nephew's analysis of a tablet from Vindonissa in Belgica: M. Alexander Speidel, "Roman Army Pay Scales," *JRS* 82 (1992): 87–106; M. Alexander Speidel, "Roman Army Pay Scales Revisited: Responses and Answers," in *De l'or pour les Braves! Soldes, Armées et Circulation Monétaire dans le Monde Romain*, ed. Michel Reddé, Collection Scripta Antiqua 69 (Bordeaux: Ausonius Éditions, 2014), 53–62.

22. Daniel Sperber, "The Costs of Living in Roman Palestine I," *Journal for the Economic and Social History of the Orient* 8 (1965): 250–251; Daniel Sperber, "The Costs of Living in Roman Palestine II," *Journal for the Economic and Social History of the Orient* 9 (1966): 190; Daniel Sperber, *Roman Palestine 200–400: Money and Prices*, Bar-Ilan Studies in Near Eastern Languages and Culture, 2nd ed. (Ramat Gan: Bar-Ilan University Press, 1991), 101–102, 122–123. Wages for women and youth, of course, were worse.

23. Josephus *A.J.* 14.417, *J.W.* 1.308; Shatzman, *Armies*, 190. Luke 3:14, while probably a creation of the evangelist, likewise states that Antipas' royal soldiers were not content with their wages.

24. Josephus *A.J.* 15.296, *J.W.* 1.403. It is not clear the extent to which the military farms in Samaria were occupied during the early Roman era (see Shimon Dar, *Landscape and Pattern: An Archaeological Survey of Samaria 800 B.C.E.–636 C.E.*, BARIS 308, 2 vols. (Oxford: BAR, 1986), 1.12–16), so one cannot be certain of the size of the Herodian military land allotments, though the deliberately poor quality of the land for purposes of farming is noteworthy.

25. Peter D. Garnsey and Greg Woolf, "Patronage of the Rural Poor in the Roman World," in *Patronage in Ancient Society*, ed. Andrew Wallace-Hadrill, Leicester-Nottingham Studies in Ancient Society (London: Routledge, 1989), 153; cf. Justin J. Meggitt, *Paul, Poverty and Survival*, Studies of the New Testament and Its World (Edinburgh: T&T Clark, 1998); Walter Scheidel, "Stratification, Deprivation and Quality of Life," in *Poverty in the Roman World*, ed. Margaret Atkins and Robin Osborne (Cambridge: Cambridge University Press, 2006), 40–59.

26. Ninety percent, including his category "stable near subsistence": Steven J. Friesen, "Paul and Economics: The Jerusalem Collection as an Alternative to Patronage," in *Paul Unbound: Other Perspectives on the Apostle*, ed. Mark Douglas Given (Peabody: Hendrickson, 2010), 37. Ninety-nine percent: Géza Alföldy, *The Social History of Rome*, trans. David Braund and Frank Pollock, Revised ed. (London: Routledge, 1988), 127; Jerry P. Toner, *Rethinking Roman History* (Cambridge: Oleander, 2002), 50–51. There was a considerable range of well-being within this spectrum. For vivid evidence of what poverty in antiquity entailed, see Gildas Hamel, *Poverty and Charity in Roman Palestine, First Three Centuries C.E.*, Near Eastern Studies 23 (Berkeley: University of California Press, 1990); Gildas Hamel, "Poverty and Charity," in *The Oxford Handbook of Jewish Daily Life in Roman Palestine*, ed. Catherine Hezser (Oxford: Oxford University Press, 2010), 308–324; Meggitt, *Paul, Poverty and Survival*, 41–74. On Palestine's economic demography in particular, see David A. Fiensy, *The Social History of Palestine in the Herodian Period: The Land Is Mine*, Studies in the Bible and Early Christianity 20 (Lewiston: Mellen, 1991), 155–176; Heinz Kreissig, *Die sozialen Zusammenhänge des judäischen Krieges: Klassen und Klassenkampf im Palästina des 1. Jahrhunderts v.u.Z.*, Schriften zur Geschichte und Kultur der Antike 1 (Berlin: Akademie, 1970), 17–87.

There were, of course, some who lived comfortable and even privileged lives in Palestine, as demonstrated in David A. Fiensy and Ralph K. Hawkins, ed., *The Galilean Economy in the Time of Jesus*, SBLECL 11 (Atlanta: SBL, 2013). However, as Sarah Rollens (Review of "Fiensy and Hawkins, eds., *The Galilean Economy in the Time of Jesus*," *Review of Biblical Literature* 2015/3 (2015): n.p.) contends, this is exactly what one would expect to find in an economically exploitative environment.

27. Raymond W. Goldsmith, "An Estimate to the Annual Structure of the National Product of the Early Roman Empire," *Review of Income and Wealth* 30 (1984): 269, acknowledging a "rather wide margin of uncertainty" with this number of working days.

28. The definitive study of Roman soldiers' diets remains Davies, *Service in the Roman Army*, 187–206. Davies' conclusions were confirmed in W. Groenman-van Waateringe, "Classical Authors and the Diet of Roman Soldiers: True or False?" in *Roman Frontier Studies 1995*, ed. W. Groenman-van Waateringe, B. L. Van Beek, W. J. H. Willems and S. L. Wynia, Oxbow Monograph Series 91 (Oxford: Oxbow, 1997), 261–266; Anthony C. King, "Animal Bones and the Dietary Identity of Military and Civilian Groups in Roman Britain, Germany and Gaul," in *Military and Civilian in Roman Britain: Cultural Relationships in a Frontier Province*, ed. T. F. C. Blagg and Anthony C. King, BARBS 136 (Oxford: BAR, 1984), 187–217.

29. Davies, *Service in the Roman Army*, 200.

30. Jonathan P. Roth, *The Logistics of the Roman Army at War (264 B.C.–A.D. 235)*, Columbia Studies in the Classical Tradition 23 (Leiden: Brill, 1999), 16–17.

31. Roman medical practice is discussed in Patricia Anne Baker, "Medical Care for the Roman Army on the Rhine, Danube and British Frontiers in the First, Second and Early Third Centuries AD" (Ph.D. Thesis, Newcastle University, 2000); Juliane C. Wilmanns, *Der Sanitätsdienst im römischen Reich: Eine sozialgeschichtliche Studie zum römischen Militärsanitätswesen neben einer Prosopographie des Sanitätspersonals*, Medizin der Antike 2 (Hildesheim: Olms-Weidmann, 1995); Davies, *Service in the Roman Army*, 209–236.

32. A medical document and a hospital were found at Masada: §20; I. A. Richmond, "The Roman Siege-Works of Masada, Israel," *JRS* 52 (1962): 162; Adolf Schulten, "Masada: Die Burg des Herodes und die Römischen Lager," *Zeitschrift des Deutschen Palästina-Vereins* 56 (1933): 127–128. Another military hospital is attested in Sebaste after the Bar Kokhba War (*BGU* 1564; 138 CE).

33. Helpful studies on the development of policies on granting auxiliaries citizenship—if somewhat out of date—include Eric Birley, "Before Diplomas, and the Claudian Reform," in *Heer und Integrationspolitik: Die römischen Militärdiplome als historische Quelle*, ed. Werner Eck and Hartmut Wolff, Passauer historische Forschungen 2 (Köln: Böhlau, 1986), 249–257; Paul A. Holder, *Studies in the Auxilia of the Roman Army from Augustus to Trajan*, BARIS 70 (Oxford: BAR, 1980), 46–63; J. C. Mann, *Britain and the Roman Empire: Collected Studies* (Brookfield: Variorum, 1996), 17–27.

34. See Sara Elise Phang, *The Marriage of Roman Soldiers (13 B.C.–A.D. 235): Law and Family in the Imperial Army*, Columbia Studies in the Classical Tradition 24 (Leiden: Brill, 2001). Contravention of this policy was taken for granted: a child born to a legionary was illegitimate and thus ineligible as an heir until their father's discharge, whereupon the lineage was legally acknowledged. Exceptions were granted to soldiers who died in active service; see the rescript of Hadrian in *BGU* 140.

35. Margaret M. Roxan, "Women on Frontiers," in *Roman Frontier Studies 1989*, ed. Valerie A. Maxfield and M. J. Dobson (Exeter: University of Exeter Press, 1991), 465.

36. Sara Elise Phang, *Roman Military Service: Ideologies of Discipline in the Late Republic and Early Principate* (Cambridge: Cambridge University Press, 2008), 19.

37. See Christopher B. Zeichmann, "Military Forces in Judaea 6–130 CE: The *status quaestionis* and Relevance for New Testament Studies," *Currents in Biblical Research* 17 (2018): forthcoming; cf. James Russell, "A Roman Military Diploma from Rough Cilicia," *Bonner Jahrbücher* 195 (1995): 67–133.

38. Legionary recruitment has been a far more popular topic among academics than that of client kings or auxiliaries. The most important works on the topic are Giovanni Forni, *Il reclutamento delle legioni da Augusto a Diocleziano*, Pubblicazioni della Facoltà di Filosofia e Lettere della Università di Pavia 5 (Milan: Bocca, 1953); Giovanni Forni, "Estrazione etnica e sociale dei soldati delle legioni nei primi tre secoli dell' Impero," *ANRW* II.1: 339–391; J. C. Mann, *Legionary Recruitment and Veteran Settlement during the Principate*, Institute of Archaeology Occasional Publication 7 (London: Institute of Archaeology, 1983). On Syria in particular, see Pollard, *Soldiers, Cities, and Civilians*, 114–134.

39. On language: Christopher B. Zeichmann, "Loanwords or Code-Switching? Latin Transliteration and the Setting of Mark's Composition," *Journal of the Jesus Movement in Its Jewish Setting* 4 (2017): 50–55; cf. extensive evidence in *DMIPERP*. On trade goods: Jodi Magness, "In the Footsteps of the Tenth Roman Legion in Judea," in *The First Jewish Revolt: Archaeology, History, and Ideology*, ed. Andrea M. Berlin and J. Andrew Overman (London: Routledge, 2002), 189–212.

40. Pollard, *Soldiers, Cities, and Civilians*, 134–138.

41. Commissioned officers in the Roman army were considerably different from commissioned officers today. The phrase refers to military offices that one could join at or above a specified rank. Today, this happens through ROTC, Military Service Academies, etc. In the Roman world, a man was eligible to be a commissioned officer not because of any specialized training, but because he held either equestrian or senatorial status. The terms are anachronistic, but useful as heuristics. In the Roman army, centurion was the highest noncommissioned office; once one became the highest ranked centurion in the legion, they were accorded equestrian status and thus eligible for commission. Most noncommissioned officers—with the exception of centurions—were termed *principales* in the Roman army.

42. §12; §38; Luke (Q) 7:1–10, Matt 8:5–13. Cf. §4; Josephus *Life* 242 on his own ordering of the Galilean rebel army. Josephus uses the Greek word *lochagos* as the royal army equivalent of *centurio*, see Hugh J. Mason, *Greek Terms for Roman Institutions: A Lexicon and*

Analysis, ASP 13 (Toronto: Hakkert, 1974), 66, 164. On centurions in the Nabataean royal army, see Graf, "Nabataean Army," 289–290.

43. Military diplomas awarded to centurions are not especially common, as most—but not all—auxiliary centurions were already Roman citizens. I am only aware of *RMD* 1.74 (*classis praetoria Antoniniana Misenensi*), *RGZM* 56 (*classis praetoria Severiana*), *AE* 2004.1913 (*cohors I Augusta c.R.*), *AE* 2005.1737 (*classis Flavia Moesia*), *AE* 2010.1852 (*cohors II Bracaraugustanorum*), *CIL* 16.103 (unit unknown), and *AE* 2014.1641 (*cohors I Lepidiana c.R.*); *RMD* 5.348 attests either a decurion or a centurion ("[ex . . .]urione"). For a (somewhat outdated) list of all known auxiliary centurions, see John Spaul, *Cohors2: The Evidence for and a Short History of the Auxiliary Infantry Units of the Imperial Roman Army*, BARIS 841 (Oxford: Archaeopress, 2000), 536–540.

44. Steve Mason, *Judean War 2: Translation and Commentary*, Brill Josephus Project 1b (Leiden: Brill, 2008), 241 n. 1925.

45. Also, Celer the auxiliary tribune (*J.W.* 2.244–246; *A.J.* 20.132, 136) and Metilius the cohort commander (*J.W.* 2.450–454). This also applies to Herodian forces: Jucundus and Tyrannus are mentioned as part of Herod's cavalry (Roman and Greek names, respectively; *J.W.* 1.527), and various other commissioned Herodian officers have Roman names including Volumnius (*J.W.* 1.535), Gratus (*J.W.* 2.58, §3), Rufus (*J.W.* 2.52, *A.J.* 17.266, §3), Crispus (*Life* 33), etc. Josephus also mentions several centurions under the prewar governor Florus as well (*J.W.* 2.319), but does not provide their names.

46. *J.W.* 2.63; Mason, *Judean War 2*, 241 n. 1925.

47. Titus Mucius Clemens: §148; S. R. Llewelyn, "The Career of T. Mucius Clemens and Its Jewish Connections," in *NewDocs* 8, 152–155. Diomedes: §§30–32. Archieus: §38.

48. On the aborted Parthian campaign: Tacitus *Ann.* 13.7. On the Jewish War: Josephus *J.W.* 2.500, 3.68.

49. Joseph Mélèze-Modrzejewski, *The Jews of Egypt: From Ramses II to Emperor Hadrian*, trans. Robert Cornman (Philadelphia: Jewish Publication Society, 1995), 185–190. The sources on Tiberius are extensive, but most important are Josephus *J.W.* 2.220, 490–97; Tacitus *Hist.* 1.11, *Ann.* 15.28; §§364–373.

50. Josephus *J.W.* 7.47–52. Antiochus' story is controverted, but I follow the reading of John M. G. Barclay, *Jews in the Mediterranean Diaspora: From Alexander to Trajan (323 BCE–117 CE)* (Edinburgh: T&T Clark, 1996), 256.

51. This paragraph draws from the fullest studies on the topic: Eric Birley, "Promotions and Transfers in the Roman Army II: The Centurionate," in *The Roman Army: Papers 1929–1986*, ed. Eric Birley, Mavors Roman Army Researches 4 (Amsterdam: Gieben, 1988), 206–220; Brian Dobson, "The Significance of the Centurion and "Primipilaris" in the Roman Army and Administration," *ANRW* II.1: 403–409; Kate Gilliver, "The Augustan Reform and the Structure of the Imperial Army," in *A Companion to the Roman Army*, ed. Paul Erdkamp, BCAW (Malden: Blackwell, 2007), 190–192; Michael P. Speidel, "Becoming a Centurion in Africa: Brave Deeds and Support of the Troops as Promotion Criteria," in *Roman Army Studies: Volume Two*, ed. Michael P. Speidel, Mavors Roman Army Researches 8 (Stuttgart: Steiner, 1992), 124–128.

52. Testimonies on auxiliary promotion are frustratingly rare, but see David Breeze, "The Career Structure below the Centurionate," *ANRW* II.1: 445–447. Breeze observes that soldiers below the rank of decurion lacked sufficient wealth to erect monuments delineating their career.

53. "*Aur(elius) Varixen, ordin(arius), qui ex fortia et suff(ragio) vex(illationis) profec(it)* . . . ." Text and translation from Speidel, "Becoming a Centurion," 124.

54. Birley, "Promotions and Transfers," 208, citing, e.g., Tacitus *Hist.* 3.49; *ILS* 2658.

2

What Did the Military Do in Early Roman Palestine?

It has become commonplace for scholars to depict the Roman military as a war machine whose purpose was to mete out physical violence on anyone threatening imperial hegemony. Prominent New Testament scholars claim that the legions "could move with all their baggage and equipment at a guaranteed fifteen miles a day to crush any rebellion anywhere," that Rome ruled foremost by means of "terror" and "humiliation," and that "Rome's ruthless and efficient military machine swept all before it . . . rebellions [were] punished with cold brutality."[1] This image is appealing and coheres with basic knowledge of the region: didn't the military engage in a major war against Judaean civilians 66–73 CE and destroy the Jerusalem temple in the process? But despite the prevalence of this image, there is much evidence that soldier-civilian interactions were usually peaceful and even positive on occasion. What does one make of this apparent contradiction? This question will be answered in two main sections. The first part will describe the "official" functions of the military in Palestine. This section will limit its attention to the duties of soldiers while on duty. The second part of this chapter will do the same with the "unofficial" functions of soldiers, examining the economic, social, and other practices common in Palestine and how they fit into Rome's broader aims.

In so doing, this chapter will address the issue of soldier-civilian interactions in Palestine and the Roman East, the topic of several recent studies.[2] The reason for limiting the scope to civilian perspectives is simple: the Gospels and Acts were written by civilians and depict events occurring in Palestine, and there is reason to suspect that at least some of the evangelists composed their Gospels there as well. This chapter thus attempts to reconstruct the sort of interactions the evangelists, their peers, and their sources may have had with the military.

EXCURSUS: PALESTINE IN ROME'S GRAND STRATEGY

One of the topics that has loomed over the study of soldier-civilian interactions in frontier zones is the issue of Roman "grand strategy." Grand strategy here refers to the role of the military in the empire's long-term objectives that interwove military, economic, and administrative policies. The modern study of Roman grand strategy was inaugurated by American military strategist Edward Luttwak.[3] The importance of Luttwak's work within the fields of Roman military, economic, frontier, and social history is difficult to overstate. Luttwak argued that from the dawn of the empire until the Flavians (27 BCE–69 CE), Roman military strategists understood the empire as four concentric circles protecting the inner zones from external threats (e.g., Parthia in the east, Germania and Caledonia in the north). 1) The innermost zone comprised Rome, Italia, and senatorial provinces, marked by the absence of military garrisons. 2) Immediately outside this zone were imperial provinces that garrisoned *auxilia* and/or legions (e.g., Syria, Egypt). Luttwak argued that these forces were garrisoned semipermanently, but were mobile forces that could move to adjacent territories should conflict arise. 3) Immediately outside Rome's provinces was its zone of diplomatic control, comprising client kingdoms (e.g., Herodian Judaea, Batanaea). These minor states maintained their own forces and acted as a cushion between Rome and neighboring regions, though imperial troops were nearby and could intervene if necessary. Client kingdoms were often immediately adjacent to perceived threats or had their own uncertainty about political stability. 4) Finally, a number of nomadic client tribes with no geographically stable state were useful because of their frequent incursions into territory at the perimeter of Roman influence and perhaps even inside the zone of their enemies' influence (e.g., the Danube region). Luttwak contended that the highly developed client system in the East compensated for the paucity of legions on that frontier: rather than risking the lives of citizen legionaries, the more disposable noncitizen royal soldiers were at stake to absorb the region's threats. Levantine military forces were primarily oriented outward against external foes, but could intervene when internal revolts occurred. The placement of legions in Figure 1.1 helps give a sense for how this worked in practice.

The literature critical of Luttwak's argument is extensive, though it sometimes devolves into caricature.[4] The numerous criticisms will not be rehearsed here, save those most pertinent to the present study. First, Luttwak tends to overlook the *internally* directed functions of the military in frontier regions. That is, Luttwak contends the central prerogative of the military was protecting the empire from external forces and so minimizes the role of the military within the empire's borders. Simply said, there is little indication that a Parthian threat figured into Palestinian military operations.[5] Rather, Josephus and other sources see virtually all military action in Palestine to be directed toward its own residents: revolt prevention or suppression, bodyguarding elites, mediating conflicts, etc. Indeed, there is no evidence that the Sebastene auxiliaries ever left Palestine before the conclusion of the Jewish War. Drawing upon insights of frontier studies, we will see that the military in Palestine did not simply protect against outside forces, but promoted various internal policies, foremost among which was facilitating the region's economic integration into the Roman Empire.

Second, there is the question of the extent to which Palestine was representative of Roman policy in general. Scholars diametrically opposed on crucial issues of the Roman military, such as Luttwak and Benjamin Isaac, nevertheless agreed that it was fairly typical of the Roman East and used it as their exemplum to construct their model of Roman Grand Strategy (or demonstrate the absence of such strategy). The Southern Levant was famous for the unrest of its civilians, banditry, and purported lawlessness—can one generalize about the Roman Empire on its basis? Scholars have increasingly seen the region as an aberration requiring distinctive policies, leading Everett Wheeler to characterize it as a military "theatre" in its own right.[6] These caveats are somewhat moot for our discussion of the Gospels, but become more relevant the farther one travels from the region.

Third, Luttwak's employment as a military strategist has an enormous impact on his writings.[7] Setting aside the fact that he has no formal degree in classics, Roman military history, or related fields, the similarities between Luttwak's description of Julio-Claudian client states and the U.S.'s strategies during the Cold War (e.g., proxy wars)—strategies that he helped devise and enact—has been disconcerting to many scholars for both reason of political-ideological implication and anachronism. Luttwak advised numerous U.S. and allied military bodies (e.g., U.S. Department of State, U.S. Navy, Army, and Air Force) and is known among other things for his open Islamophobia and promoting the odious notion that we should "give war a chance." His scholarly work should be read with these proclivities in mind, as his writings have a normative element that authorize very specific—and problematic—foreign policies.

Though not directly related to Luttwak, one must name the elephant in the room regarding the military and the Levant: the present-day Israel-Palestine conflict. This conflict has been a common subtext in research of the military in the Levant. One particularly obvious example can be found in the scholarship of Mordechai Gichon.[8] Gichon begins his retrospective on the study of the Judaean frontier with the long history of conflicts between Jews and their neighbors to the east, then discusses the Jewish Herodian government's policing and fortification policies, whether Rome maintained a border defense to protect Judaea, the southern extent of Judaean territory, and the persistent threat of unprovoked attacks from stateless "Arabs" upon peaceful Jewish civilians—matters that correspond with talking points about Israel-Palestine.

OFFICIAL FUNCTIONS OF THE MILITARY IN PALESTINE

The single best known function of the military was its role in combat. Palestine was subjected to numerous military interventions and combat missions from the invasion of Pompey (63 BCE) through the Bar Kokhba War (132–136 CE). Combat was primarily directed against foes internal to Palestine (e.g., Hyrcanus II vs. Aristobulus II, post-Herod revolts, Jewish War, Bar Kokhba War), though there were a few incidents against external forces (e.g., Antipas vs. the Nabataean king Aretas IV, Agrippa II aiding Rome against Parthia). Military combat occurred with some regularity in Palestine, but was usually localized. Most conflicts, even those directed against Palestinian civilians, would have no immediate implications for the vast majority of other Palestinian residents: for instance, Pilate's slaughter of a Samaritan group at Mount Gerizim probably had little effect upon other residents of Samaria. Even the region-wide combat of the Jewish War left many sites unaffected by destruction. The size and diversity of Palestine meant that it was generally easy for civilians to avoid participation in military conflicts if their village was not directly implicated. Most soldiers garrisoned in Palestine never went to battle, with the obvious exception of those serving 66–73 CE.[9]

This might be contrasted with how the military functioned in Jewish and Palestinian memory: literature written in or about early Roman Palestine tends to remember combat vividly and associates the military with traumatic events (e.g., Ps. Sol. 17, War Scroll, As. Mos. 6–7, Sib. Or. 1–5, Josephus). The Gospels are similarly interested in soldiers mostly insofar as they harm civilians. Despite its centrality to the passion narrative, evidence of soldiers meting out punishments on civilians is sparse. Physical punishments such as scourging, beating, and beheading were usually the prerogative of lictors, not soldiers. In prewar Palestine, however, governors did not hold sufficient status to have lictors as bodyguards, so it is conceivable that soldiers meted out such punishments. Scholarly knowledge of soldiers' role in criminal discipline during the early Roman Empire is extremely limited. When punishment by beating is described in ancient literature, those meting it out are rarely mentioned, if simply because the punishers were rarely important as characters.[10] Rather, texts attribute the beatings to the superiors who order it, such as magistrates and governors.

Military forces also carried out preventative violence to deter unrest. Examples from Palestine are numerous, but Felix's attack upon an Egyptian prophet is representative: the governor allegedly heard rumors of the prophet's intent to attack Jerusalem and decided it better to strike first, successfully dissipating the prophet's following (Josephus *J.W.* 2.261–263; *A.J.* 20.169–171). Roman and Herodian policies of preventative violence has been a topic of interest among social historians recently, largely focused on state violence against bandits (*lēstai*). This vein of scholarship has emphasized the phenomenon of "social banditry," which sees a connection between the guerilla tactics of socially marginal groups as operating alongside a form of economic banditry. New Testament scholars sometimes depict bandits as Robin Hood–like figures who took from the rich and protected poor villagers, but there is reason to doubt this was the case in Judaea. Josephus notes that Herod forced these brigands out of Sepphoris, but the bandits then terrorized residents of surrounding villages; this hardly sounds like a populist movement against Romanized elite. It is probable that Herod's military intervention was even welcomed by locals, since it brought an end to their harassment. In another instance, after the Jewish War, the residents of Ascalon likely dedicated a monument to a centurion named Aulus Instuleius Tenax for aiding the city after Jewish rebels sacked the city in 68 CE (§9; cf. *J.W.* 2.460, 2.477). Herodian military colonies were founded in Batanaea at least partially because of the region's reputation for outlaws.[11]

Another important duty was patrol routines. There tend to be two rather extreme understandings of military patrols and resulting interventions into civilian affairs.[12] On the one hand, rabbinic sources depict army patrols as systematic and unwanted bands of foreigners interrupting daily life in Jewish villages. Indeed, rabbis imply that soldiers sexually assaulted Jewish women and abused civilians whenever opportunity arose. Consequently, the rabbis maneuvered Torah to permit abrogation should the army arrive unexpectedly and threaten the well-being of Jewish villagers.[13] Benjamin Isaac points to a story from a second-century source: "A patrol of gentiles came into town and [the townspeople] were afraid that [the soldiers] might harm them and therefore we prepared them a calf and we fed them and gave them drink and rubbed them with oil so that they would not harm the townspeople."[14] Here the rabbis describe an exception to the rule that no food should be prepared for Gentiles on festival days; the mere presence of soldiers was potentially dangerous enough that typical norms did not apply. Patrolling soldiers' hostility is also apparent in Josephus' works: one auxiliary soldier destroyed a village's Torah scroll (*A.J.* 20.115, *J.W.* 2.229) and another "flashed" temple attendees (*A.J.* 20.108, *J.W.* 2.224).

On the other hand, military intervention was actively sought by distressed civilians.[15] Dozens of Egyptian documents attest civilians petitioning military officers for protection or to bring offenders to justice. One papyrus representative of the genre reads as follows:

> To Longinus, decurion of the Arsinoite nome, from Pakebkis son of Onnophris from the village of Tebtunis, exempted priest of the famous temple in the village. On the 30th of the month Epeiph, when the hour was late, one Saturnilus, with a great many others, I know not why, having no complaint against us, picked a quarrel, going so far as to rush in with staves, and seizing my brother Onnophris they wounded him, so that his life is endangered in consequence. Wherefore, sir, being careful for the danger to his life, I submit this statement and beg you to order [the perpetrator] to be brought before you so that he may take the consequences, and that I may obtain the requisite satisfaction. . . . (*P.Tebt.* 2.304)

The inverse of the rabbis' depiction, wherein civilians feared a roving gang of army thugs, holding out no hope for recompense, Pakebkis understands the military as an outlet for justice against malignant peers. About half of these petitionary papyri seek redress for physical assault, indicating that Pakebkis was hardly alone in this experience. Petitions sometimes took the form of a request for protection against agitated parties; one letter pleas that a centurion might prevent violence by an unsuccessful litigant that became disgruntled (*SB* 5238). Though most petitioners claim "some kind of special status," many widows and other marginalized civilians tugged at the heartstrings of the district centurion, drawing attention to their humble state.[16] Those against whom action is hoped to be taken are extremely diverse: relatives, mobs, thieves, and even town elders.

We can deduce that the soldier-civilian relations reflected in Egyptian texts were found in Palestine as well. First, Egyptian documents attest these practices over the duration of the early Roman period. Over sixty papyri attest this phenomenon from 20 BCE to 255 CE. Thus, even though Egypt was subject to its own political vicissitudes, the content of these petitions remains strikingly consistent over a long period of time. Second, Josephus attests a few high-profile civilian petitions to imperial officers acting in a military capacity: the legate Petronius interceded on behalf of Jewish delegation to halt Gaius' order to install a statue in Jerusalem (*A.J.* 18.269–278, *J.W.* 2.192), the governor Cumanus was petitioned to bring justice during a Galilean-Samaritan conflict (*A.J.* 20.119, *J.W.* 2.233), etc. Josephus has no interest in the mundanities of village life, but one of the few exceptions to this silence is his comment that Vespasian set up decurions in villages and centurions in towns during the Jewish War to facilitate the rebuilding of both social and physical structures (*J.W.* 4.442).

Third, papyri from a later period indicate that officers garrisoned in Syria Palaestina (i.e., third-century Palestine) also received petitions to intervene on behalf of locals, including a Jewish woman who sought justice for her brother's murder.[17] Similar petitions have been found among tablets in British Vindolanda, indicating the practice was spread across the empire.[18] At least one civilian inscription commends the virtue of a military officer in the Jewish War (§9; cf. §10, §160). While the rabbinic depiction of the Roman military is mostly negative, the Mishnah and Talmud recount several positive anecdotes—for instance, soldiers reportedly supervised Jewish legal studies at Jamnia in the school's formative days.[19]

The fourth and most compelling reason to suppose that the Egyptian and Palestinian policing were similar is that these contradictory patterns of military policing—acting as both bully and protector—occurred simultaneously in Egypt itself. A papyrus so absurd it verges on comedy is a private accounting ledger containing multiple and escalating entries for money extorted by soldiers among more typical expenses (*SB* 9207). It begins:

> To the guard on duty: 2 drachmae, 1 obol
> Gift: 240 drachmae
> Suckling pig: 24 drachmae
> To the guard: 20 drachmae . . .

Perhaps most ridiculous are two subsequent entries in the same ledger: one explicitly labeled "extortion" (*hyper diaseismou*) at 2,200 drachmas, with another at 400 drachmas "to the soldier by demand"! The record also includes payouts to the police chief: apparently the unlucky author had once attempted to bring an end to strongarmed offers of "protection" by appealing to the soldiers' superior through the sort of petitions mentioned earlier. This appeal was unsuccessful, given the aforementioned payout and 400 drachmas subsequently extorted by yet another soldier. Entries for extortion combine for the greatest expenses in the ledger by a wide margin. The mere fact that these were now listed as *expenses* indicates that the author had lost any confidence in eventual recompense; he had literally written the money off as a loss.[20] A more optimistic example can be found in another Egyptian papyrus. The author, a civilian named Hermon, sought redress against both soldiers and civilians for the theft of fish from his pond and subsequent threats of violence.[21] Hermon asked the centurion to pursue recompense for the fish and to prevent further harassment; we do not know if he was successful. These papyri recall the scene at the Jordan River in Luke 3:14: "The soldiers also asked [John the Baptist], 'And what should we do?' He said to them, 'Do not extort (*diaseisēte*) money from anyone by threats or false accusation, and be satisfied with your wage.'" While this tradition probably reflects Luke's compositional context more than Herodian Palestine, it nevertheless suggests a widespread pattern of interaction in the Roman East.

Soldiers' duties on patrol were directly related to the geography of garrison and fortification placement. Nearly all Judaean wars and insurrections strategically concerned the control of fortifications, so Herodian kings spent large sums fortifying existing cities for greater security.[22] While the use of major fortresses was significant for urban dwellers, more notable for rural inhabitants were minor forts, watchtowers, and road stations dispersed along the provincial road system. Literary sources rarely remark on their existence, but many such structures have been uncovered by archaeologists and are noted in Figure 2.1. The clearest evidence comes from the Jaffa-Jerusalem roads. Surviving evidence indicates several military structures during the Herodian and post-annexation period. The distribution of these towers and fortlets was not proportionate to the density of nearby civilian settlements, though military structures tended to cluster within ten kilometers of fortified cities at satellite villages (e.g., three or four west of Jerusalem, another one or two near Emmaus). These small fortified villages facilitated active policing and were proximate to major fortified cities suggests in order to aid intercommunication in the event of trouble. Egyptian ostraca document two guards of low rank at each watchtower: one spending the day up in the tower and the other patrolling the road;[23] presumably Palestinian towers were manned similarly.

One might expect increases in patrolling and fortifications after the Jewish War; the influx of soldiers, new roads, and escalated tensions could have led to a greater density of village patrols that kept civilians from considering further revolt. This, somewhat counterintuitively, was not the case. Instead, most rural towers and forts were abandoned around the time of the war. Though there were nine prewar military sites on the Jaffa-Jerusalem roads, there were only five after the War and all but one of them functioned as satellites of the Jerusalem garrison.[24] Israel Shatzman makes a persuasive case for the systematic abandonment of military sites in the Negev after the Jewish War: most likely Beersheba, Tel 'Ira, Tel Sharuhen, Tel 'Arad, and Tel 'Aro-

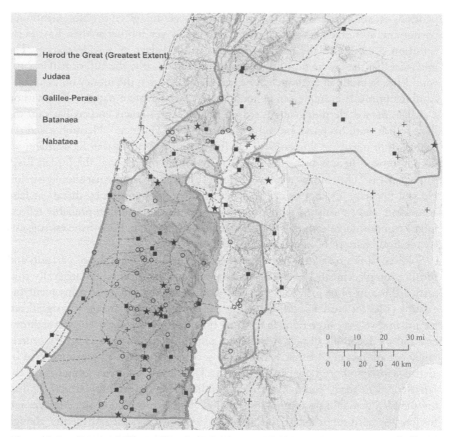

Figure 2.1.   Roman Military Sites in Palestine, 63 BCE–132 CE. A cross (+) indicates either a site where there is no reason to suspect military presence, but a relevant find was discovered, or an item from *DMIPERP* that was discovered outside of Palestine. An empty circle (○) indicates a site where there was probably only a prewar military presence. A solid square (■) indicates a site where there was probably a military presence both before and after the war. A solid pentagram (★) indicates a site where there was probably only a postwar military presence.

rer were vacated after the war, leaving only a handful of isolated sites along Judaea's southern border.[25] Many other military sites were deserted after the Jewish War, as can be seen in Figure 2.1. Even the rural sites that did have a military presence after the war tended to see a reduction in number of soldiers (e.g., Masada, Jericho).[26] This policy of abandonment may be attributed to a number of causes: patrolling was a kindness that Judaean civilians no longer deserved, refocusing strategy to the protection of economically and strategically important sites, the Flavian shift to unified policies of large garrisons across the empire, etc. The Decapolis, by contrast, had little military presence before or after the war—David Kennedy and others note that there is no evidence of rural garrison in the region and most urban sites evince a limited military presence.[27]

Because patrolling declined after the Jewish War, soldiers were more heavily concentrated at large garrisons. While the exact size of the prewar Jerusalem garrison is not certain, the presence of *legio X Fretensis* alone was sufficient to double the size of the entire prewar Judaean army. Similarly, *legio VI Ferrata* garrisoned at Caparcotna/Legio and its environs starting 120 CE. Two auxiliary units, *cohors I Thracum miliaria* and *ala Antiana Gallorum et Thracum sagittaria*, garrisoned near Beth Guvrin at Khirbet 'Arak Hala.[28] Other examples could be cited. Benjamin Isaac suggests that the prewar distribution of soldiers in small villages to aid civilians thinned out forces in a potentially disastrous manner, as had happened in Germania under Varus' command in 9 CE.[29] Given the tensions in Roman-Jewish relations after the war, it was deemed better to prepare for the worst than to attempt repair of the relationship. Legionaries were thus concentrated in massive garrisons with far fewer soldiers in minor sites; should another revolt break out, Rome would control access to financial resources, food stores, strategic sites, etc.

These smaller detachments were placed to address concerns other than patrolling and policing. The Healing of the Centurion's Slave (Q/Luke 7:1–10, Matt 8:5–13), for instance, is plausible in its depiction of an officer in Capernaum. Capernaum had a complex history of administration, but it was usually a border town from 4 BCE until the Bar Kokhba War ended in 135 CE and thus likely to have bureaucrats in need of protection (see Table 2.1). Indeed, Mark 2:13–14 depicts Levi the tax-collector as stationed on the shore of the Lake of Gennesaret at Capernaum. Egyptian papyri indicate that tax collectors and customs officials regularly encountered or caused problems for locals and are named in the sort of petitions described above. Two contrasting examples will suffice.[30] On the one hand, centurions might intervene on behalf of civilians who were abused by tax collectors: one plea comes from a family unable to pay the full grain tax, leading a tax collector and his companions to strip and beat the family matriarch to the point of severe injury. One of her sons sought justice against the tax collector's gang through the centurion. On the other hand, a petition from a tax collector beseeches the aid of a decurion against local hunters. The hunters not only reported their income fraudulently and refused to comply with tax laws, but started abusing the tax collector and others. The connection between military officers and collection of tolls, taxes, etc., is also evidenced in the client kingdom of Nabataea: a first-century text designed to aid sailors comments that a centurion and his men aided a customs official exacting tolls at an Nabataean port.[31] A centurion placed near Capernaum could ensure the protection of those extracting tolls and taxes from merchants on regional trade routes, or perhaps those exacting a fish tax (cf. *OGIS* 496), either of which would explain Levi's presence at the border/coastal village of Capernaum.

Table 2.1. The Military in Palestine 66 BCE–135 CE, Focusing on Capernaum

	Judaea	Galilee	Batanaea
66 BCE Civil war	IK. Aristobulus II: king. RA; aid from mercenaries.		IK. Ptolemy son of Mennaeus: king. RA.
63 BCE Pompey's conquest	CK. Hyrcanus II: ethnarch. RA; aid from Rome during conquest.		CK. Ptolemy son of Mennaeus: tetrarch. RA.
57 BCE 5 Synods	Synods. Synedria at Jerusalem and Jericho. Legions.	Synod. Synedrion at Sepphoris. Legions.	
47 BCE Return to monarchy	CK. Hyrcanus II: ethnarch; Antipater: procurator. RA.	CK. Hyrcanus II: ethnarch; Herod: governor. RA.	
44 BCE Antipater dies	CK. Hyrcanus II: ethnarch; Phasael: governor. RA.		
42 BCE	CK. Hyrcanus II: ethnarch; Phasael: tetrarch. RA.	CK. Hyrcanus II: ethnarch; Herod: tetrarch. RA.	
40 BCE Uprising	IK. Antigonus: king. RA; aid from Parthia and Batanaea during conquest.		CK. Lysanias: tetrarch. RA.
37 BCE	CK. Herod: king. RA; aid from Rome during conquest.		CK. Cleopatra: queen; Zenodorus: lessee. RA.
30 BCE			CK. Zenodorus: tetrarch. RA.
23 BCE Batanaea split	CK. Herod: king. RA.		
4 BCE Herod dies	CK. Archelaus: ethnarch. RA: inherited most of Herod's RA; aid from legions and Nabataea during revolts.	CK. Antipas: tetrarch. RA: inherited few of Herod's RA.	CK. Philip: tetrarch. RA: inherited few of Herod's RA.
6 CE Archelaus ousted	EP. Coponius: prefect. Local *auxilia*: formed from Archelaus' RA.		
34 CE Philip dies	EP. Pilate: prefect. Judaean *auxilia*.		Syria: Consular Roman province. Vitellius: legate. Legions and *auxilia*.
37 CE	EP. Marullus: prefect. Judaean *auxilia*.		CK. Agrippa I: king. RA.
39 CE Antipas banished		CK. Agrippa I: king. RA.	

41 CE Judaea reunified	CK. Agrippa I: king. RA: formed from Judaean *auxilia*.		
44 CE Agrippa I dies	EP. Fadus: procurator. Judaean *auxilia*: formed from Agrippa I's army.		
53 CE	EP. Felix: procurator. Judaean *auxilia*.		CK. Agrippa II: king. RA.
55 CE Batanaea expanded	EP. Felix: procurator. Judaean *auxilia*.		CK. Agrippa II: king. RA.
66 CE Galilee revolts	EP. Antonius Julianus: procurator. Legions and various *auxilia*.	IK. Josephus: governor. Galilean rebels.	CK. Agrippa II: king. RA.
67 CE Galilee subdued	EP. Marcus Antonius Julianus: legate. Legions and various *auxilia*; aid from Batanaea, Commagene, and Emesa.		
70 CE Judaea promoted	Praetorian Roman province. Sextus Vettulenus Cerialis: legate. Legions and foreign *auxilia*.		CK. Agrippa II: king. RA.
96 CE Agrippa II dies	Praetorian Roman province. Sextus Hermidius Campanus: legate. One legion and foreign *auxilia*.		Syria: Consular Roman province. Lucius Junius Caesennius Paetus: legate. Legions and *auxilia*.
120 CE Judaea promoted	Consular Roman province. Lucius Cossonius Gallus: legate. Two legions and foreign *auxilia*.		
132 CE Bar Kokhba War	Israel: IK. Simon Bar Kokhba: prince. RA.	Judaea: Consular Roman province. Quintus Tineius Rufus: legate. Legions and foreign *auxilia*.	
135 CE Roman victory	Syria Palaestina: Consular Roman province. Cnaeus Minicius Faustinus Sextus Iulius Severus: legate. Legions and foreign *auxilia*.		

[Administrative status]. [Head of military]: [his title]. [Troops present]: [troops' origination, if noteworthy];
   [external aid, if any].
Abbreviations: CK: client kingdom allied with Rome. EP: equestrian subprovince of Roman Syria. IK: independent kingdom not allied with Rome. RA: royal army.
Shaded cell indicates the kingdom or province governing Capernaum.

Soldiers also constructed public works. Two major civic building projects are notable in Palestine. First, road paving was a significant infrastructural effort with heavy military involvement; the earliest evidence of military road construction in

Palestine is a milestone on the Caesarea-Scythopolis road erected during the Jewish War (§297), though a far more substantial paving effort occurred during the reign of Hadrian in both Judaea (§§302–315) and the Decapolis (§§336–339). William Arnal observes that road pavement not only expedited travel for merchants and the military, but also facilitated tax collectors' journeys to villages that had been inaccessible.[32] Settlements that had previously been isolated were now integrated into the provincial and imperial economy. Consequently, villages that were small-but-self-sufficient during the Herodian period came to focus crop production on offering resources to nearby Palestinian cities. Rather than farming various crops or producing multiple products for communities in the immediate vicinity, farmers and craftspeople focused their efforts on the needs of larger markets of the closest city.[33] The effects of this economic shift were mostly negative, leading to extensive debt.

The other major civic construction projects were waterworks such as aqueducts and wells. Judaea had aqueducts since the beginning of the early Hellenistic era; by the end of the Herodian period, there were aqueducts to over a dozen Palestinian cities, some supplied by multiple systems.[34] Our interest begins with the annexation of Judaea in 6 CE, since Rome tended to involve soldiers in such construction efforts.[35] Of the Judaean aqueducts built during the first century, Josephus only sees fit to mention the construction that Pilate oversaw in Jerusalem, since he had intended to fund it with money from the temple treasury (*A.J.* 18.60–62, *J.W.* 2.175–177; cf. Luke 13:1–5). Despite Jewish outcry about the funding, this work was eventually completed, now known as Jerusalem's 'Arrub aqueduct. Though Josephus is not interested in who constructed these aqueducts, we find clearer evidence about a century later under Hadrian. Hadrian financed aqueducts elsewhere in the empire at several million sestertii each, as part of a large infrastructural effort; a number of inscriptions indicate that legionaries constructed Hadrian's aqueduct to Caesarea (§§121–130; see Figure 2.2).[36]

Beyond these civic constructions, the army built structures for their own purposes. Soldiers produced brick tiles for these constructions, with the surplus sold to civilians for private use.[37] Such bricks are easily identified by the stamps bearing the unit's name and sometimes its symbols (see Figure 3.2). Civilian attitudes about military construction were complex, as evident in a story from the Mishnah:

> Rabbi Judah, Rabbi Jose, and Rabbi Simeon were sitting, and Judah, a son of proselytes, was sitting near them. Rabbi Judah commenced [the discussion] by observing, "How fine are the works of [the Romans]! They have made streets, they have built bridges, and they have erected bath houses." Rabbi Jose was silent. Rabbi Simeon bar Yohai answered and said, "All that they made they made for themselves; they built marketplaces, to set prostitutes in them; baths, to rejuvenate themselves; bridges, to levy tolls for them."[38]

One is reminded of the famous scene in *Monty Python's Life of Brian* (1979): "But apart from better sanitation, and medicine, and education, and irrigation, and public health, and roads, and a freshwater system, and baths, and public order—what have the Romans done for us?!"

Figure 2.2. Hadrian's Aqueduct in Judaea. This aqueduct to Caesarea Maritima was erected as part of the emperor Hadrian's massive construction efforts in the Roman East (117–138 CE). Several legions worked together to build this particular aqueduct: *legiones II Traiana, VI Ferrata, X Fretensis,* and *XXII Deiotariana* (see §§121–130).
Photo credit: Oren Rozen, used by permission.

Unofficial Functions of the Military in Palestine

If we were to limit our discussion to the on-duty responsibilities of soldiers, we would have an incomplete portrait of their role in Palestine. Much of their role in the region extended beyond their explicit duties to more subtle purposes. To start with the most obvious, the military was heavily involved in the economy of Palestine through purchase of supplies. Because Palestine's economy was relatively isolated until Hadrian's reign, items were purchased locally. Josephus records an incident where a bandit gang slaughtered forty soldiers and a centurion at Emmaus; these soldiers were supplying their garrison with corn and weapons (*A.J.* 17.282–283; 4 BCE). Also revealing is a legionary expense sheet from Masada, which lists some of the items the soldier purchased at the market: barley, a linen tunic, boots, and a broad category of "food expenses" (§22; cf. §149). The Roman economy had little resemblance to free-market capitalism, so the military dictated its own prices—for instance, papyri from Dura-Europos indicates that an auxiliary cohort paid merely 125 denars for each of its replacement horses, regardless of their age or quality.[39] The limited evidence from Palestine suggests prices tended toward fairness rather than extortion, but presumably both occurred.[40]

As in other provinces, markets (*canabae* and *vici*) emerged around more permanent Palestinian army bases, evident in the archaeological record.[41] Cassius Dio states that Palestinian Jews even sold weapons to the Roman army until the outbreak of the Bar Kokhba War (69.12.2). In other cases, merchants might form part of the army supply train that followed a unit around. Lucius Tettius Crescens (§190) may have been one such merchant. Though his epitaph intentionally misleads the reader so as to infer he was a soldier, Miriam Pucci Ben Zeev argues that he was likely a Roman businessman (perhaps a slave trader) who followed the army into Judaea.[42] It is not certain how these transactions were perceived by Palestinian civilians, though graffiti at Khirbet 'Arak Hala suggests some held a positive view of these interactions: soldiers purchased olive oil from civilians (§69) and an Aramaic inscription at the site reads, "May the memory of Lord Trajan be blessed . . ." (§75). By way of contrast, *O.Ber.* 2.126 (61 CE) is a letter from an Egyptian merchant who refused to work at the military's contract rates, deeming them extortion.

Around the time of Jesus' death, there were approximately 7,000 soldiers in Palestine, each in need of life's requirements. This number increased to roughly 13,000 after the Jewish War—now mostly under the authority of a single administrative apparatus as opposed to distinct kingdoms and provinces.[43] This additional population in itself would amplify the economic importance of the military, but three factors rendered the army even more central to the economy after the Jewish War.

First, soldiers helped monetize the Palestinian economy. Before the war, coinage was uncommon, such that taxes were exacted in kind (e.g., grain, animal produce) rather than via cash.[44] The extent of monetization—that is, the use of coin for transaction—during the prewar period was uneven. Rural areas evince a disproportionately large number of low-value coins, suggesting that locals preferred to barter; by contrast, Jerusalem, Caesarea, and much of Judaea proper were far more monetized. Cities already tended to be economic magnets, siphoning agricultural and other material surpluses from nearby villages. This is not to suggest that money was entirely alien to rural areas. Rather, coinage was unnecessary for many Palestinians, thanks to other methods of economic transaction such as trade.

Soldiers' purchase of their own goods introduced coinage into exchange processes that had previously occurred via barter or payment in kind. The introduction of legions led to a massive influx of imperial coinage. The Roman Empire had a confusing set of competing monetary systems: provincial coinage which was legal tender only in one specific province or kingdom (and often adjacent ones) and imperial coinage which was usable anywhere within the empire; the New Testament mentions coins within both of these systems: Judaean provincial coinage (e.g., lepton Mark 12:42), Syrian provincial coinage (e.g., drachma Matt 15:8, didrachma Matt 17:24), and Roman imperial coinage (e.g., quadrans Mark 12:42, denarius Mark 12:5). Since Roman imperial coins were not always available for payment, legions often stamped provincial coins with a countermark to legally render them imperial coinage, a practice especially well attested by the legions of Palestine.[45] By merely injecting imperial coins into the economy, the military substantially changed the character of the Judaean economy, especially in rural areas; even in those areas thoroughly monetized, there was a shift from provincial to imperial coinage after the war. As Danny Syon describes it, "in [the post-War period], however, things change dramatically. All over the southern Levant, the number of *denars* found increased nearly fourfold and the *aes* coins sixfold. . . . It is quite clear that Roman imperial coins started arriving in Palestine in any appreciable numbers only in the Flavian period, i.e., from around 70 CE."[46] Coinage provided far easier means of extracting and transporting taxes, rent, loans, tithes, bribes, and extortion than the cumbersome use of material goods.[47] Like elsewhere, the extent of monetization correlates directly with the prominence of the area's military presence.[48] In short, more soldiers meant greater use of coinage. Local taxes and imperial tributes were the primary means of funding troop pay, rendering the process cyclical in that more coinage resulted in easier exaction of taxes.

Second, the Jerusalem garrison of *legio X Fretensis* partially filled the economic vacuum left by the temple's destruction during the Jewish War. As noted above, goods and money tended to flow toward cities, but Jerusalem was the primary economic motor of Judaea during the prewar period thanks to its massive temple apparatus.[49] In addition to the collection of tithes, Jerusalem housed a sizable landowning aristocracy, required sacrificial victims, and saw the regular influx of Jewish pilgrims. Jerusalem's population required more resources than its immediate vicinity could furnish, so they drew from a much larger area. The temple had long been *the* central economic institution in Palestine and its destruction left a massive void. Though there had long been a garrison in Jerusalem, the economic role of the army was considerably greater after the war due to the increase in number of soldiers, the higher pay of legionaries than auxiliaries, paving roads connecting Jerusalem to nearby cities (thus easing tax collection and trade), the presence of foreign legionaries with standardized purchasing habits, and the shift toward city-based garrisons. Even the sale of sacrificial animals was partially replaced by the meat-heavy diet of soldiers. Jonathan Roth notes that while the Jerusalem garrison "in no way substituted for the Temple as an economic motor," the military eliminated "many of the constraints that the Temple and its priesthood had placed on the Judaean economy and had an overall positive effect on its growth in the long term."[50] While Roth's point is important, the "positive effect" should be disputed; Jeffrey L. Davies instead suggests that Rome's "military demands . . . consistently drained the agricultural resources of a locality rather than enhanced it."[51] Regardless, the army aided the economic expansion of Jerusalem and acted as the default replacement for the temple's commercial activity.

Third, soldiers and veterans became more heavily involved in landownership in Palestine after the war, entailing extensive interaction with civilians through their *personal* financial transactions. The emperor Vespasian confiscated Judaean land that had belonged to rebels and their sympathizers, granting this property to favored citizens, including veterans and loyalists. Josephus records the gift of property in Emmaus to a settlement of 800 veterans (*J.W.* 7.217) and archaeological and epigraphic finds suggest veterans may have been settled at other sites as well (see Figure 2.1). These settlements incorporated veterans into existing Palestinian communities, which also had the effect of integrating these smaller locales into province-wide structures. The abundance of property and land grants positioned veterans as a strong presence among the landed and they could rent out their own property to locals as private citizens. Rabbinic midrashim attest Jewish tenancy on land owned by military officers in the late first century, land which had been owned by Jewish civilians before the war.[52]

The increased use of coinage in the Palestinian economy entailed a sharp increase in debt among small landholders, thanks to the newfound ease of financial extraction.[53] The upsurge of debt is attributable to an increase in loans. The best attestation to this phenomenon in Palestine is a document attesting a forty denars loan from the centurion Magonius Valens to a Jewish villager named Judah in Ein-Gedi (§43). The surviving document is a Greek copy of an Aramaic original treating a courtyard as collateral for a loan. Magonius was part of a temporary garrison (*cohors I Thracum miliaria*) that camped on both the eastern and western border of the courtyard in question, as well as a *praesidium* to the north; a bathhouse in the area was discovered and may have served this cohort.[54] The document was given to Judah, while the Aramaic original stayed with Magonius, leading to a strange situation where neither the creditor nor the debtor held a copy in a language they found intelligible;[55] the witnesses include a combination of Jewish and Roman names—the latter names presumably belonging to other auxiliaries. In this particular case, it seems the centurion convinced Judah to sign an agreement to repay a sixty denars loan, when in fact Judah had only received forty: in the initial copy of the text, the word forty is scratched out and replaced by sixty, but only in the second copy did the scribe correctly list the amount Magonius demanded.[56] A later papyrus from the same archive (*P.Yadin* 19) indicates that Judah managed to repay the loan, since he still owned the courtyard a few years later.[57] Other documents attesting loans between soldiers and Palestinian or Jewish civilians have been discovered (§36, §58). Soldiers also acted as business partners to provincials: a centurion named Lucius initially ensured protection of an Egyptian tax collector Nemesion, but eventually Nemesion helped Lucius circumvent laws against soldiers owning land in their garrison province by partaking in shared agricultural enterprises.[58]

While soldiers' pay was hardly exorbitant, they were among the fortunate few to have surplus income. Their habits of consumption—especially after the temple's fall—comprised a major part of the Romanization of Palestine. Roman-style theaters are evinced at likely garrison and fortress sites, including Antipatris, Sebaste, Jerusalem, Jericho, Caesarea, and Gerasa.[59] Bath houses were another common means of catering to the Roman tastes of legionaries and veterans.[60] While these buildings could be partially attributed to the tastes of populations already residing in these cities, evidence from the village of Caparcotna is only explainable by the legionary presence. Caparcotna had a theater to entertain the nearby garrison of *legio VI Ferrata*, even though the village was otherwise unremarkable. Indeed, the Caparcotna came to be the nexus of *three* Judaean roads and was renamed Legio on account of its legion. Evidence is similar near Beth Guvrin: the village was barely populated during the first century, but with the postwar garrison it steadily grew in prominence; an amphitheater was built there after the Bar Kokhba War and it eventually received "city" status as Eleutheropolis in 200 CE. Likewise, Shechem was reestablished as the veteran colony Neapolis in 72 CE, where numerous entertainment buildings were found. Indeed, it is nearly a rule that military fortresses were accompanied by amphitheaters throughout the empire.[61] Jonathan Roth notes that officers' considerably higher wages "would have created a market not only for the best wine and foodstuffs, but for other luxury goods as well."[62] Even common legionaries would bring with them the tastes and habits of consumption that their life as a Syrian and Roman citizen had cultivated.

That said, merchants did not simply import goods to market to legionaries after the war. Before the Jewish War, Roman influence on locally produced wares was minimal.[63] At larger prewar garrisons (e.g., Jerusalem, Caesarea, Herodium), one finds distinctively Roman wares imported from elsewhere, such as lamps and *fibulae*—albeit with little influence on other goods. This was probably because Palestine's armies were relatively autonomous before the Jewish War (e.g., empowered to fight their own wars like those against the Nabataeans, but nevertheless subservient to Rome), most detachments consisted of a few men per fortlet, and their soldiers had long used goods similar to civilians and continued doing so without interruption. Once Syrian legionaries arrived, local craftspeople adapted to the new tastes, marking a change in the wares locals produced. Somewhat surprisingly, pottery and similar goods were imported less frequently in the postwar period.

This pattern of auxiliaries' imported goods that had little influence upon local wares and legionaries' Romanization of local comestibles is attested elsewhere on the Roman frontier. Gregg Woolf notes that in early Roman Gaul, Roman influence on consumption habits tended to be small goods that announced the owner's new Roman persona—such as objects of adornment that were also common in prewar garrisons.[64] These wares were inexpensive and mundane, but signaled a transformation of noncitizen auxiliaries into a new Romanized person, albeit within the soldiers' poorly informed conception of what constituted Roman consumption, in anticipation of their eventual citizenship. By contrast, citizen legionaries were less interested in imported permanent wares for several reasons, not least of which was a markedly different conception of Roman consumption habits and little need for conspicuous Romanized consumption. One central distinction seems to be auxiliaries' appreciation of Roman objects as opposed to legionaries' preference for objects of Roman appreciation. That is, auxiliaries preferred goods that conferred "Romanness" upon them in a manner visible to others (e.g., clothing, ornaments), whereas legionaries tended to prefer mundane objects that may have reminded them of their more-Roman home (e.g., lamps, pottery).

Low-level officers also acted as financial benefactors to Palestinian communities—men with sufficient wealth to permit generosity, but low-ranking enough that they regularly interacted with provincial civilians. Nigel Pollard suggests that their distance from major sites of Roman power allowed these officers—usually centurions and decurions—to exert influence they otherwise would not have.[65] The villagers in

Phaena (formerly in the kingdom of Batanaea) honor a legionary centurion named Petusius Eudemus as friend and benefactor (*ton philon kai euergetēn*; *IGRR* 3.1122; cf. 3.1121). A contemporary centurion also in former Batanaean territory was thanked by the residents of Aere for similar reasons.[66] An inscription of Titus Flavius Dionysius may also attest to this phenomenon in Gerasa (§33).

Finally, family life is well attested.[67] Several Palestinian epitaphs were erected on behalf of soldiers or veterans by their families (§5, §35, §53, §148; cf. §147, §§198–200), including spouses, siblings, children, and freedpersons. During the period of their military service, soldiers commonly corresponded with their families. Most of the data survives from Egypt, including letters from an anxious brother (*P.Mich.* 8.484), siblings petitioning officers to get their brother transferred to a closer garrison (*P.Oxy.* 1666), a soldier seeking his father's approval for the purchase of a slave woman (*P.Mich.* 8.476), among many others. A particularly vivid letter comes from a soldier named Saturnilus stationed in Nubia to his mother Aphrodous in Egyptian Karanis (*P.Mich.* 3.203; 114–116 CE). Saturnilus related a variety of concerns to his mother, foremost among them is the birth of another child. Aphrodous sent her son supplies regularly (olives and other unnamed allowances) and had also been tending pigs intended for her grandchildren. Saturnilus had been collecting letters as an excuse to travel to Karanis to see his mother, but the presence of the prefect has made him cautious, lest Saturnilus' explanation prove inadequate and his supplies be confiscated. Another example is a strange divorce contract indicating a peculiar position the military could put a family in: a couple involuntarily divorced because the husband enlisted in the army; only 400 of the 1,000 drachmas dowry were returned in the divorce, probably indicating the couple planned to remarry upon his discharge.[68] In prewar Palestine, auxiliary and client kings' soldiers rarely left the province, but the postwar garrison of *legio X Fretensis* was recruited primarily from Syria. Brief leaves of absence would permit legionaries to visit home on occasion. It is worth noting that even after the War, some veterans opted to remain in Palestine and make it their new home (§41, §§53–55, §148)—just as some Palestinian veterans opted not to return to their home region (§§199–200, §204, §209, §257, §263, §§293–296).[69]

CONCLUSION

As far as military-civilian interactions are concerned, the history of Rome in Palestine can be divided into three eras. The first extends from the conquest of Pompey to the start of the Jewish War (63 BCE–66 CE). During this period, soldiers were recruited from within the province and a slow process of Romanization occurred within the army. There was also a similarly steady process of de-Judaizing of the military's demographic as well, but the military's role should be characterized as an army of *local garrison* with the primary duty of policing. The second phase extends from the end of the Jewish War to the Bar Kokhba War (73–132 CE). This period is marked by a sudden increase in the size of the military and its spread within the region of Palestine following their destruction of the temple. Moreover, soldiers were no longer recruited locally and their foreignness becomes a detectable theme in literature of the period. In short, the military presence became one of *occupation*, having recently completed conquest of Judaea. Third, after the Bar Kokhba War (135 CE), the territory became emphatically Roman in a manner distinct from even the previous period. That is, Palestine had become sufficiently Romanized through *settler colonialism* that the entire province was no longer thought of as a site of occupation—residual Jewish and Palestinian sovereignty was eliminated, and Jews were largely relegated to Galilee.[70]

Through all of this, it is clear that the model of the "total institution," largely separate from the surrounding culture and self-sufficient, is unhelpful for understanding the role of the Roman army in Palestine either before or after the Jewish War.[71] Rather, we see the military was a major facet of daily life—both positive and negative—for denizens of Palestine.

NOTES

1. John Dominic Crossan, *God and Empire: Jesus against Rome, Then and Now* (San Francisco: HarperOne, 2007), 12; Richard A. Horsley, *Jesus and Empire: The Kingdom of God and the New World Disorder* (Minneapolis: Fortress, 2003), 27–31; N. T. Wright, *Paul: In Fresh Perspective* (Minneapolis: Fortress, 2009), 64. It should be noted that Roman historians tend not to hold this view, instead seeing the empire more positively, though this is increasingly moderated.

2. The first major publication on the topic was Ramsay MacMullen, *Soldier and Civilian in the Later Roman Empire* (Cambridge: Harvard University Press, 1966). See also, e.g., Richard Alston, *Soldier and Society in Roman Egypt: A Social History* (London: Routledge, 1995); T. F. C. Blagg and Anthony C. King, ed., *Military and Civilian in Roman Britain: Cultural Relationships in a Frontier Province*, BARBS 136 (Oxford: BAR, 1984); Benjamin Isaac, *The Limits of Empire: The Roman Army in the East*, Revised ed. (Oxford: Clarendon, 1992); Pollard, *Soldiers, Cities, and Civilians*.

3. Edward N. Luttwak, *The Grand Strategy of the Roman Empire from the First Century A.D. to the Third*, 1st ed. (Baltimore: Johns Hopkins University Press, 1976).

4. On Luttwak's deeply flawed understanding of the siege of Masada, the central example of his monograph, see Hannah M. Cotton, "The Impact of the Roman Army in the Province of Judaea/Syria Palaestina," in *The Impact of the Roman Army (200 BC–AD 476): Economic, Social, Political, Religious, and Cultural Aspects*, ed. Lukas De Blois and Elio Lo Cascio, Impact of Empire 6 (Leuven: Brill, 2006), 400–406. Among the most important works relating to Roman grand strategy, especially with respect to the East: Warwick Ball, *Rome in the East: The Transformation of an Empire* (London: Routledge, 2000); Stephen Dyson, *The Creation of the Roman Frontier* (Princeton: Princeton University Press, 1985); Hugh Elton, *Frontiers of the Roman Empire* (Bloomington: Indiana University Press, 1996); Arther Ferrill, *Roman Imperial Grand Strategy*, Publications of the Association of Ancient Historians 3 (Lanham: University Press of America, 1991); Isaac, *Limits of Empire*; Kimberly Kagan, "Redefining Roman Grand

Strategy," *Journal of Military History* 70 (2006): 333–362; David L. Kennedy, *The Roman Army in Jordan*, 2nd ed. (London: Council for British Research in the Levant, 2004); J. C. Mann, "Power, Force and the Frontiers of the Empire," *JRS* 69 (1979): 175–183; Susan P. Mattern, *Rome and the Enemy: Imperial Strategy in the Principate* (Berkeley: University of California Press, 1999); Fergus Millar, "Emperors, Frontiers and Foreign Relations, 31 B.C. to A.D. 378," *Britannia* 13 (1982): 1–23; Fergus Millar, *The Roman Near East: 37 BC–AD 337*, Carl Newell Jackson Lectures (Cambridge: Harvard University Press, 1993); Maurice Sartre, *The Middle East under Rome*, trans. Catherine Porter and Elizabeth Rawlings (Cambridge: Harvard University Press, 2005); Adrian N. Sherwin-White, *Roman Foreign Policy in the East* (Norman: University of Oklahoma Press, 1984); Everett Wheeler, "Methodological Limits and the Mirage of Roman Strategy," *Journal of Military History* 57 (1993): 7–41, 215–240; C. R. Whittaker, *Frontiers of the Roman Empire: A Social and Economic Study*, Ancient Society and History (London: Johns Hopkins University Press, 1994). See also the proceedings for the Congress of Roman Frontier Studies and similar conferences.

5. Isaac, *Limits of Empire*, 22. But see the objections in Wheeler, "Methodological Limits," 19–20. The best sustained arguments against a primarily outward-directed military in the East (i.e., defending from Parthia) can be found in Isaac, *Limits of Empire*; Pollard, *Soldiers, Cities, and Civilians*; Whittaker, *Frontiers of the Roman Empire*.

6. E.g., Everett Wheeler, "The Army and the *Limes* in the East," in *A Companion to the Roman Army*, ed. Paul Erdkamp, BCAW (London: Blackwell, 2007), 235–366.

7. Edward N. Luttwak, "Give War a Chance," *Foreign Affairs* 78/4 (1999): 36–44. Luttwak released an "updated and revised" edition of *Grand Strategy*, which has not fared well in scholarly opinion; reviews have been negative, citing *inter alia* his unwillingness to learn from his critics, whom he dismisses as lacking "any experience of military or policy planning, let alone war, unlike the present writer. . . ." (*The Grand Strategy of the Roman Empire from the First Century A.D. to the Third*, Revised and Updated ed. [Baltimore: Johns Hopkins University Press, 2016], xii).

8. Mordechai Gichon, "45 Years of Research on the *Limes Palaestinae*: The Findings and Their Assessment in the Light of the Criticisms Raised (C1st–C4th)," in *Limes XVIII: Proceedings of the XVIIIth International Congress of Roman Frontier Studies*, ed. Philip Freeman, Julian Bennett, Zbigniew T. Fiema and Birgitta Hoffmann, BARIS 1084, 2 vols. (Oxford: BAR, 2002), 1.185–206.

9. One gets the impression of restless boredom from soldiers' graffiti at Herodium before the Jewish War (§§117–120).

10. E.g., Cicero *Verr.* 2.5.161–162; Aulus Gellius *Noct. att.* 10.3.1–20; Suetonius *Aug.* 45.4; Philo *Flacc.* 78–80; John 19:1; Acts 16:22–23; 2 Cor 11:24–25.

11. For this policy in general, see Peter Richardson, *Herod: King of the Jews and Friend of the Romans*, Studies on Personalities of the New Testament (Edinburgh: T&T Clark, 1996), 139–142, 232, though Richardson later revised this position, arguing that Judaean watchtowers were *not* placed to keep bandits in check (*Building Jewish in the Roman East*, Supplements to the Journal for the Study of Judaism 92 (Leiden: Brill, 2004), 24–25). Note also that the career of Aulus Instuleius Tenax is controverted, contrast §§9–10 and *CIIP* 2335.

12. On policing the Roman provinces, see Christopher J. Fuhrmann, *Policing the Roman Empire: Soldiers, Administration, and Public Order* (Oxford: Oxford University Press, 2012), 172–200.

13. Benjamin Isaac, "The Roman Army in Judaea: Police Duties and Taxation," in *Roman Frontier Studies 1989*, ed. Valerie A. Maxfield and M. J. Dobson (Exeter: University of Exeter

Press, 1991), 458–461; Isaac, *Limits of Empire*, 115–118. Similar interactions are related in m.'Abod.Zar. 5.6; 'Erub. 3.5. Eusebius *Hist. eccl.* 8.12.3 reports likewise with Antioch.

14. T.Beṣah 2.6, translation from Isaac, "Roman Army in Judaea," 458.

15. Alston, *Soldier and Society*, 81–96; Jean-Jacques Aubert, "Policing the Countryside: Soldiers and Civilians in Egyptian Villages in the 3rd and 4th Centuries A.D.," in *La hiérarchie (Rangordnung) de l'armée romaine sous le Haut-Empire*, ed. Yann Le Bohec, De l'archéologie à l'histoire (Paris: De Boccard, 1995), 257–265; Davies, *Service in the Roman Army*, 175–185. See Alston, *Soldier and Society*, 87–91 for a helpful overview of surviving Egyptian evidence of petitions to military officers. Updated overviews can be found in John Whitehorne, "Petitions to the Centurion: A Question of Locality?" *Bulletin of the American Society of Papyrologists* 41 (2004): 161–169; Michael Peachin, "Petition to a Centurion from the NYU Papyrus Collection and the Question of Informal Adjudication Performed by Centurions (*P.Sijp.* 15)," in *Papyri in Memory of P. J. Sijpesteijn (P.Sijp.)*, ed. A. J. B. Sirks and K. Worp, ASP 40 (Oakville: ASP, 2007), 86–88.

16. Alston, *Soldier and Society*, 91–92.

17. *SB* 15496–15500, see especially 15497 and 15500, the latter of which is mentioned here and dated to 234 CE.

18. *T.Vindol.* 257, 281, 322, 344; Michael Peachin, "Five Vindolanda Tablets, Soldiers, and the Law," *Tyche* 14 (1999): 223–235. Whitehorne, "Petitions to the Centurion" argues that the geographical lopsidedness of the papyrological record is attributable to the demands of frontier regions and attendant inaccessibility of higher authorities for legal recourse (e.g., strategoi, prefects).

19. y.B.Qam. 4.4a, so Shimon Applebaum, "Judaea as a Roman Province: The Countryside as a Political and Economic Factor," *ANRW* II.8: 395; cf. Ze'ev Safrai, "The Roman Army in the Galilee," in *The Galilee in Late Antiquity*, ed. Lee I. Levine (New York: Jewish Theological Seminary of America, 1992), 113, citing t.Šabb. 13.9; y.Šabb. 15d; y.Yoma 8.5.45b; y.Ned. 4.9.38d; b.Šabb. 121a.

20. Caesar's decree that money not be taken from Judaeans cannot be seriously considered as reflective of actual practice, so it presumably occurred there as well. Josephus *A.J.* 14.20: *mēde stratiōtais exē chrēmata toutōn eisprattesthai*; cf. *A.J.* 14.195.

21. *P.Oxy.* 2234 (31 CE); cf. *BGU* 4, 908; *SB* 5280.

22. Benjamin Isaac, "Roman Administration and Urbanization," in *Greece and Rome in Eretz Israel: Collected Essays*, ed. Aryeh Kasher, Uriel Rappaport and Gideon Fuks (Jerusalem: Israel Exploration Society, 1990), 151–159; Duane W. Roller, *The Building Program of Herod the Great* (Berkeley: University of California Press, 1998). Cf. Figure 2.1.

23. Alston, *Soldier and Society*, 81–82, citing *O.Amst.* 8–14.

24. Moshe Fischer, Benjamin Isaac, and Israel Roll, *Roman Roads in Judaea II: The Jaffa-Jerusalem Roads*, BARIS 628 (Oxford: Tempvs Reparatvm, 1996).

25. Israel Shatzman ("The Beginning of the Roman Defensive System in Judaea," *American Journal of Ancient History* 8 [1983]: 130–160; *Armies*, 233–246) offers a devastating critique of Mordechai Gichon's arguments to the contrary (e.g., Gichon, "45 Years"). Gichon developed the earlier thesis of Albrecht Alt ("Limes Palaestinae," *Palästinajahrbuch* 26 (1930): 43–82) that, after the Jewish War (Gichon: under Vespasian; Alt: under Trajan), a defensive perimeter was developed on the southern Judaea-Nabataea border—that is, a *Limes Palaestinae*. Gichon contends that this system had its origins in a highly developed fortification system created by Herod the Great. There are significant problems with Gichon's dating of numerous sites, inference of a road connecting Raphia to the Dead Sea, and the extent of

Judaea's southern limits. See similar critiques of Gichon's thesis in M. H. Gracey, "The Armies of Judean Client Kings," in *Defence of the Roman and Byzantine East*, ed. David L. Kennedy and Philip Freeman, BARIS 297, 2 vols. (Oxford: BAR, 1986), 1.311–318; Adam Pažout, "Spatial Analysis of Early Roman Fortifications in Northern Negev" (Diploma thesis, Charles University in Prague, 2015); Yoram Tsafrir, "The Desert Fortresses of Judaea in the Second Temple Period," in *The Jerusalem Cathedra: Studies in the History, Archaeology, Geography and Ethnography of the Land of Israel*, ed. Lee I. Levine, 3 vols. (Detroit: Wayne State University Press, 1982), 2.120–145.

26. It is worth noting as a caveat that soldiers also billeted in archaeologically indistinct buildings. For instance, no specifically military structures have been found in postwar Emmaus and the existing fortifications were abandoned after the war's conclusion. Nevertheless, inscriptional data (§§27–29, §59) renders an Emmaus garrison beyond doubt: Moshe Fischer and his co-authors contend that the epigraphic evidence indicates a sufficient stability of military presence for a stonemason to set up shop. Fischer, Isaac, and Roll, *Roman Roads in Judaea II*, 151–160.

27. Lamia el-Khouri, "The Roman Countryside in North-west Jordan (63 BC–AD 324)," *Levant* 40 (2008): 73; Kennedy, *Roman Army in Jordan*, 112–114.

28. Boaz Zissu and Avner Ecker, "A Roman Military Fort North of Bet Guvrin/Eleutheropolis," *ZPE* 188 (2014): 293–312.

29. Benjamin Isaac, "Reflections on the Roman Army in the East," in *Defence of the Roman and Byzantine East*, ed. David L. Kennedy and Philip Freeman, BARIS 297, 2 vols. (Oxford: BAR, 1986), 2.389–390; Isaac, *Limits of Empire*, 107. Isaac cites Cassius Dio 56.19.1–5 on Varus and Battle of the Teutoburg Forest.

30. *BGU* 515 (193 CE) and *PSI* 3.222 (late III CE), respectively; cf. *P.Corn.* 90; *P.Mich.* 7.425, 10.582; *BGU* 81; *P.Gen.* 1.17; *SB* 9203; *P.Oxy.* 1185; *P.Amh.* 2.77. A. E. Hanson ("Village Officials at Philadelphia: A Model of Romanization in the Julio-Claudian Period," in *Egitto e storia antica dall'ellenismo all'età araba: Bilancio di un confronto*, ed. L. Crisculo and G. Geraci [Bologna: Cooperativa Libriria Universitaria Editrice Bologna, 1989], 435–436; "Sworn Declaration to Agents from the Centurion Cattius Catullus: P.Col. Inv. 90 [*P.Thomas* 5]," in *Essays and Texts in Honor of J. David Thomas*, ed. Traianos Gagos and Roger S. Bagnall, ASP 42 [Oakville: ASP, 2001], 91–97) discusses the archive of a first-century collector of capitation taxes in Egyptian Philadelphia and its evidence for military bodyguards.

31. Peripl. M. Rubr. 19. It is not clear whether the centurion is Nabataean or Roman; see the contrasting arguments in Isaac, *Limits of Empire*, 125; Pollard, *Soldiers, Cities, and Civilians*, 101. On evidence for similar activities in rabbinic Palestine, see Rab. (Lev) 30:6 and Isaac, *Limits of Empire*, 282 n. 98. See also the assessment of Capernaum's role in regional trade routes in James H. Charlesworth and Mordechai Aviam, "Reconstructing First-Century Galilee: Reflections on Ten Major Problems," in *Jesus Research: New Methodologies and Perceptions*, ed. James H. Charlesworth, Princeton-Prague Symposia Series on the Historical Jesus 2 (Grand Rapids: Eerdmans, 2014), 124–127; Jonathan L. Reed, *Archaeology and the Galilean Jesus: A Re-examination of the Evidence* (Harrisburg: Trinity Press International, 2000), 146–148.

32. William E. Arnal, *Jesus and the Village Scribes: Galilean Conflicts and the Setting of Q* (Minneapolis: Fortress, 2001), 115–155. Benjamin Isaac is careful to note, however, that "Roman roads were constructed for the use of the military organization in the provinces. The economic benefits for the local provincial population, which may have resulted from their existence, were a by-product rather than a primary aim in their construction." ("Infrastruc-

ture," in *The Oxford Handbook of Jewish Daily Life in Roman Palestine*, ed. Catherine Hezser [Oxford: Oxford University Press, 2010], 147.)

33. Jonathan P. Roth, "The Army and the Economy in Judaea and Palaestina," in *The Roman Army and the Economy*, ed. Paul Erdkamp (Amsterdam: Gieben, 2002), 389–390. On the open-closed economy question in Palestine, see Arnal, *Jesus and the Village Scribes*, 115–133; Ze'ev Safrai, *The Economy of Roman Palestine* (London: Routledge, 1994), 415–435.

34. See the overview in Joseph Patrich and David Amit, "The Aqueducts of Israel: An Introduction," in *The Aqueducts of Israel*, ed. David Amit, Joseph Patrich and Yizhar Hirschfeld, JRASup 46 (Portsmouth: JRA, 2002), 9–20.

35. P. A. Février, "L'armée romaine et la construction des aqueducs," *Dossiers de l'archéologie* 38 (1979): 88–93. Note the presence of a military architect in Syria Palaestina (*AE* 1983.380).

36. Leah Di Segni, "The Water Supply of Roman and Byzantine Palestine in Literary and Epigraphical Sources," in *The Aqueducts of Israel*, ed. David Amit, Joseph Patrich and Yizhar Hirschfeld, JRASup 46 (Portsmouth: JRA, 2002), 54–55; Ramsay MacMullen, "Roman Imperial Building in the Provinces," *Harvard Studies in Classical Philology* 64 (1959): 210.

37. Renate Rosenthal-Heginbottom, "The Material Culture of the Roman Army," in *The Great Revolt in the Galilee*, ed. Ofra Guri-Rimon, Haifa Museum 28 (Haifa: University of Haifa, 2008), 93–94.

38. b.Šabb. 33b. The rabbis of the anecdote lived during the second century.

39. *P.Dura* 56 (208 CE); *P.Dura* 97 (251 CE).

40. E.g., Gaius Messius (§22) paid seven denars for a linen tunic, which is more than the eight drachma (two denars) in first- and early second-century Egypt (*P.Oxy.* 285 in 50 CE, which is extortive; *P.Oxy.* 1269 in early II CE). Gaius paid five denars for boots, which is comparable to military purchases at sixteen drachma (roughly four denars) in Egypt (*CPL* 106 in 81 CE).

41. Safrai, "Roman Army in Galilee," 110–114; Safrai, *Economy of Roman Palestine*, 346. On *canabae*, see César Carreras Monfort, "The Roman Military Supply during the Principate: Transportation and Staples," in *The Roman Army and The Economy*, ed. Paul Erdkamp (Amsterdam: Gieben, 2002), 70–89; Jeffrey L. Davies, "Soldiers, Peasants and Markets in Wales and the Marches," in *Military and Civilian in Roman Britain: Cultural Relationships in a Frontier Province*, ed. T. F. C. Blagg and Anthony C. King, BARBS 136 (Oxford: BAR, 1984), 129–142; Jeffrey L. Davies, "Native Producers and Roman Consumers: The Mechanisms of Military Supply in Wales from Claudius to Theodosius," in *Roman Frontier Studies 1995*, ed. W. Groenman-van Waateringe, B. L. Van Beek, W. J. H. Willems and S. L. Wyrria, Oxbow Monograph Series 91 (Oxford: Oxbow, 1997), 267–272; Roth, *Logistics of the Roman Army*, 96–101.

42. Miriam Pucci Ben Zeev, "L. Tettius Crescens' *Expeditio Iudaea*," *ZPE* 133 (2000): 256–258.

43. In the Herodian period, soldiers include those in Judaea (3,400 auxiliaries), Ascalon (600 auxiliaries), Batanaea (2,000 soldiers?), Galilee (1,000 soldiers?), plus various bodyguards and mercenaries. Postwar soldiers include Judaea (4,800 legionaries, 3,400 auxiliaries), Batanaea (4,000 soldiers?), Ascalon (600 soldiers), plus other irregulars; see Jonathan P. Roth, "The Length of the Siege of Masada," *SCI* 14 (1995): 91–92. Legionaries doubled in number under Trajan (Werner Eck, "Zum konsularen Status von Judaea im frühen 2. Jh.," *Bulletin for the American Society of Papyrologists* 22 [1984]: 55–67; Isaac, *Limits of Empire*, 105–107), though Batanaea was divided between Judaea and Syria by that point. Numbers would be considerably higher if the Decapolis were included.

44. Fabian E. Udoh, *To Caesar What Is Caesar's: Tribute, Taxes, and Imperial Administration in Early Roman Palestine 63 BCE–70 CE*, Brown Judaic Studies 343 (Providence: Brown Judaic Studies, 2005), 221–223.

45. The countermarks of *legio X Fretensis* in the East are uniquely prevalent among all legionary countermarks (see Christopher Howgego, *Greek Imperial Countermarks: Studies in the Provincial Coinage of the Roman Empire*, Royal Numismatic Society Special Publication 17 (London: Royal Numismatic Society, 1985), 22 [with nos. 117, 281–282, 291, 409–410, 727–735]; Mayer Rosenberger, *The Coinage of Eastern Palestine and Legionary Countermarks, Bar-Kochba Overstrucks* (Jerusalem: Rosenberger, 1978), 78–84). By Howgego's (now outdated) count, an astounding 71 percent of all legionary countermarks recovered throughout the empire were stamped by *legio X Fretensis*!

46. Danny Syon, *Small Change in Hellenistic-Roman Galilee: The Evidence from Numismatic Site Finds as a Tool for Historical Reconstruction*, Numismatic Studies and Researches 11 (Jerusalem: Israel Numismatic Society, 2015), 212.

47. The process of monetization through urbanization in Palestine is discussed by several experts, but I adopt a model similar to that of Arnal, *Jesus and the Village Scribes*, 134–150. 1) Tax collection was difficult in rural Palestine, resulting in a de facto exemption in these areas. 2) Increased urbanization (as bureaucratic centers), monetization (as a means of transporting wealth), and new infrastructure (increasing accessibility) made tax collection considerably easier. 3) The resulting loss of income among villagers led to an increase in debt and loss of property. 4) To forestall debt, villagers often produced specialized goods that maximized profit. 5) Specialization required villagers to shift away from self-sufficient subsistence-farming and toward cash-cropping, resulting in their economic dependence on cities and elites. Objections to Arnal's model concern the particulars of Antipas' Galilee, especially the absence of silver coinage most commonly used in tax exaction (e.g., Mark A. Chancey, *Greco-Roman Culture and the Galilee of Jesus*, SNTMS 134 [Cambridge: Cambridge University Press, 2005], 181; Morten Hørning Jensen, *Herod Antipas in Galilee: The Literary and Archaeological Sources on the Reign of Herod Antipas and Its Socio-Economic Impact on Galilee*, WUNT II 215, 2nd ed. [Tübingen: Mohr Siebeck, 2010], 190–191). That is, objections tend more toward the application of the model to the dataset of Herodian Galilee than the validity of the model itself. I apply the model to the postwar period, a period wherein the sharp increase in the production and use of silver currency is unambiguous. Consequently, such criticisms are less pertinent to the present study. Moreover, it is clear from the military pay receipts (§22, §36) that soldiers tended to use imperial coinage. The most relevant models for monetization, coinage, and the military for first-century Palestine can be found in Constantina Katsari, "The Monetization of the Roman Frontier Provinces," in *The Monetary Systems of the Greeks and Romans*, ed. W. V. Harris (Oxford: Oxford University Press, 2008), 242–267; Pollard, *Soldiers, Cities, and Civilians*, 171–211; David G. Wigg, "Coin Supply and the Roman Army," in *Roman Frontier Studies 1995*, ed. W. Groenman-van Waateringe, B. L. van Beek, W. J. H. Willems and S. L. Wyrria, Oxbow Monograph Series 91 (Oxford: Oxbow, 1997), 281–288.

48. On the empire broadly, see Michael Crawford, "Money and Exchange in the Roman World," *JRS* 60 (1970): 46. In Egypt, denars are found foremost in military contexts until their formal introduction into the economy in 296 CE (Erik Christiansen, "On Denarii and Other Coin-Terms in the Papyri," *ZPE* 54 [1984]: 271–299). On Palestinian cities, see Syon, *Small Change*, 212–213; Roth, "Army and the Economy," 384 in turn citing data from Crawford, "Money and Exchange," 43.

49. See the discussion in Hayim Lapin, "Jerusalem the Consumer City: Temple, Cult, and Consumption in the Second Temple Period," in *Expressions of Cult in the Southern Levant in the Greco-Roman Period: Manifestations in Text and Material Culture*, ed. Oren Tal and Zeev Weiss, Contextualizing the Sacred 6 (Turnhout: Brepols, 2017), 241–254.

50. Roth, "Army and the Economy," 391. On the temple's role in regulating the local economy, Roth cites Safrai, *Economy of Roman Palestine*, 37. Cf. Oliver Stoll, *Zwischen Integration und Abgrenzung: Die Religion des Römischen Heeres im Nahen Osten. Studien zum Verhältnis von Armee und Zivilbevölkerung im römischen Syrien und in Nachbargebieten*, Mainzer Althistorische Studien 3 (St. Katharinen: Scripta Mercaturae, 2001), 380–417.

51. Davies, "Native Producers," 271–272.

52. Midr. Sifre Devei Rav 317; on the dating and interpretation, see Applebaum, "Judaea as a Roman Province," 389–392. Applebaum cites §148 and §§54–55 as instances of rural veteran colonization in Palestine.

53. See Arnal, *Jesus and the Village Scribes*, 134–150.

54. Hannah M. Cotton, "Courtyard(s) in Ein-gedi: *P.Yadin* 11, 19 and 20 of the Babatha Archive," *ZPE* 112 (1996): 197–201; Hannah M. Cotton, "Ein Gedi Between the Two Revolts," *SCI* 20 (2001): 148–149. Cotton also observes that comparison with *P.Yadin* 19 indicates the cohort left Ein-Gedi by April 128, but §211 shows it remained in Palestine. However, Gwyn Davies and Jodi Magness recently disputed that the cohort ever garrisoned in Ein-Gedi ("Was a Roman Cohort Stationed at Ein Gedi?" *SCI* 32 [2013]: 195–199).

55. Jacobine G. Oudshoorn, *Roman and Local Law in the Babatha and Salome Komaise Archives: General Analysis and Three Case Studies on Law of Succession, Guardianship and Marriage*, Studies on the Texts of the Desert of Judah 69 (Leiden: Brill, 2007), 156.

56. Naphtali Lewis, *The Documents from the Bar-Kochba Period in the Cave of Letters: Greek Papyri*, Judean Desert Studies 2 (Jerusalem: Israel Exploration Society, 1989), 41; Oudshoorn, *Roman and Local Law*, 160.

57. See Cotton, "Courtyard(s) in Ein-gedi" for a discussion of how the courtyards mentioned in §43, *P.Yadin* 19 and 20 relate to each other.

58. Hanson, "Sworn Declaration," 94–95.

59. Arthur Segal, *Theatres in Roman Palestine and Provincia Arabia*, MnemosyneSup 140 (Leiden: Brill, 1994); Zeev Weiss, *Public Spectacles in Roman and Late Antique Palestine*, Revealing Antiquity 21 (Cambridge: Harvard University Press, 2014) discuss all known Palestinian theaters, hippodromes, and amphitheaters. On evidence for military presence at these sites, see the gazetteer in *DMIPERP*.

60. See Haynes, *Blood of the Provinces*, 171–174.

61. Davies, *Service in the Roman Army*, 67.

62. Roth, "Army and the Economy," 286.

63. Cotton, "Impact of the Roman Army"; Moshe Fischer, "Rome and Judaea during the First Century CE: A Strange *modus vivendi*," in *Fines imperii—imperium sine fine?* ed. Günther Moosbaur and Rainer Wiegels, Osnabrücker Forschungen zu Altertum und Antike-Rezeption 14 (Rahden: Leidorf, 2011), 143–156.

64. Greg Woolf, *Becoming Roman: The Origins of Provincial Civilization in Gaul* (Cambridge: Cambridge University Press, 1998), 169–205. Note that this was not always the case; the material cultures of auxiliaries and legionaries stationed in Germania Inferior, for instance, are indistinguishable.

65. Pollard, *Soldiers, Cities, and Civilians*, 88. Pollard documents ample evidence of the phenomenon in Syria. Note also two cases of veteran officers from *cohors I Damascenorum* becoming a priest of Serapis in Oxyrhynchus (§§194–195).

66. *ILS* 2413. See other inscriptions of centurions as benefactors for temples listed in Graeme A. Ward, "Centurions: The Practice of Roman Officership" (Ph.D. Dissertation, University of North Carolina at Chapel Hill, 2012), 224 n. 31.

67. See the fullest treatment of soldiers and families: Phang, *Marriage of Soldiers*; cf. Pollard, *Soldiers, Cities, and Civilians*, 151–159.

68. *P.Mich.* 7.422; Robert O. Fink, "*P.Mich.* VII 422 (Inv. 4703): Betrothal, Marriage, or Divorce?" in *Essays in Honor of C. Bradford Welles*, ed. Alan E. Samuel, ASP 1 (New Haven: ASP, 1966), 9–17.

69. To be sure, the vast majority of legionaries stationed in the Levant returned to their home province. Mann (*Legionary Recruitment*, 150) counts twenty-eight of thirty legionaries (69–160 CE) in Eastern provinces returning to their home outside the province of their garrison upon retirement, though these numbers are severely out of date. Veterans in the west seem to have strongly favored staying in the province of their garrison, by contrast.

70. This schema overlaps with but deviates from Ze'ev Safrai's tripartite schema of military presence in Palestine. See Safrai, *Economy of Roman Palestine*, 339; Safrai, "Roman Army in the Galilee," 104, which identify the transfer of *legio VI Ferrata* in 120 CE as the pivotal moment introducing the third phase in that the number of soldiers probably increased from 10,000 to around 25,000 and the province saw a change in administrative status. Cf. Chancey, *Greco-Roman Culture*, 43–70.

71. *Contra* T. R. Hobbs, "Soldiers in the Gospels: A Neglected Agent," in *Social-Scientific Models for Interpreting the Bible: Essays by the Context Group in Honor of Bruce J. Malina*, ed. John J. Pilch, Biblical Interpretation 53 (Leiden: Brill, 2001), 328–348; Nigel Pollard, "The Roman Army as a "Total Institution" in the Near East? Dura-Europos as a Case Study," in *The Roman Army in the East*, ed. David L. Kennedy, JRASup 18 (Ann Arbor: JRA, 1996), 211–228; Brent D. Shaw, "Soldiers and Society: The Army in Numidia," *Opus* 2 (1983): 133–159.

3

The Military in the Gospels and Acts

The military is a major force in the canonical Gospels and Acts: soldiers execute Jesus, Jewish newborns, John the Baptist, and several Christians. Despite these acts of cruelty, soldiers are often depicted positively: the centurion Cornelius is the first Gentile Christian and the kindly tribune Claudius Lysias ensures Paul's protection from hostile mobs in Jerusalem. What allows these wildly divergent portraits to exist alongside one another? To what extent can and should they be reconciled with one another?

This chapter will discuss how knowledge of military matters—matters like those discussed in the previous chapters—can elucidate the narratives and politics of the canonical Gospels and Acts. Attention will be limited to the most explicit invocations of the military and will omit discussion of other interesting, albeit tangential, topics like spiritual warfare, the politics of the "Kingdom of God," and the role of the military in the economic world of these narratives. Over the course of this chapter we will see that even though the Gospels often tell the same or similar stories, the significance of these narratives differs with each text. As such, this chapter will attend to both the *distinctive voice* of each evangelist and the significance of that voice in the *author's compositional context*: how Mark, for instance, depicts the military and why he depicted the military the way he did at this particular time and location of his writing. Greater attention will be devoted to Mark, since this book operates on the assumption that the other three canonical evangelists knew and used Mark in writing their own Gospels; thus, by understanding the nuances of Mark, we can better see how the later evangelists made use of and modified what they found in Mark.

THE GOSPEL OF MARK

Mark, the earliest of the surviving Gospels, was likely composed shortly after the Jewish War somewhere in Galilee, perhaps Capernaum.[1] William Arnal has persuasively argued that the author of Mark is a refugee of the Jewish War, having fled from Jerusalem to safer land in Galilee.[2] The trauma of the war lingers beneath the surface of Mark's Gospel, with Jesus often facilitating replacement of the Jerusalem temple—offering the forgiveness, gathering place, and ritual activities that were formerly associated with the temple. Even though the author of Mark was a Jewish Palestinian, he does not remember Jerusalem fondly, mostly depicting it as a place of depravity. We will see that the war informs many narratives where soldiers appear, as the evangelist even welcomes the temple's destruction.

The Gerasene Demoniac—Mark 5:1–20

The Exorcism of the Gerasene Demoniac is often interpreted as polemic against the Roman army.[3] There are two primary arguments supporting a military subtext in the passage. First, the demonic force is named Legion (5:9; 5:15). *legio* is among Mark's occasional Latinisms; in other instances where Mark transliterates Latin terms into Greek, it almost uniformly signals something encoded as Roman in the postwar period (e.g., *denarius, centurio, praetorium*).[4] In other ancient texts, the word *legio* always refers to a Roman military unit, which Mark presumably evokes as well. Legions were a significant aspect of the social landscape in Palestine after the Jewish War, so this reference would have been particularly notable in Mark's context. Second, Mark uses images associated with the specific legion garrisoned in postwar Judaea, *legio X Fretensis*. This legion was active throughout Judaea during the Jewish War and after the war garrisoned in the city of Jerusalem, which it helped destroy. The symbols of *legio X Fretensis* included a boar, a ship, and the deity Neptune who was god of the sea, each of which Mark may evoke in his telling of the story: Jesus sends Legion into pigs that drown in the Sea of Galilee. Mark probably knew these symbols, as they were prominent in postwar Palestine; for example, a boar and war galley are featured on roof tiles produced by *legio X Fretensis* (see Figure 3.2), Neptune is found on some of its inscriptions (see Figure 3.1), and the legion used images of pigs, boats, and dolphins to countermark Judaean coins. This is not to mention the presence of a boar on the ceremonial signum for *legio X Fretensis* itself. Consequently, Judaean residents were likely familiar with the legion's imagery and may have associated such symbols with *legio X Fretensis* in particular.[5]

The transliteration of *legio* is sufficient to suggest some kind of military component to the pericope; the cumulative weight of *legio X Fretensis*' symbols confirm the matter beyond doubt for some interpreters. Commentators almost uniformly understand this pericope's allusion to the army as polemic against the Roman occupation of Palestine: "Legion" is not only demonic, but also evokes the ritually unclean presence of pigs. This anti-occupation reading sometimes takes on an allegorical quality:

Figure 3.1. Neptune on an Inscription of *legio X Fretensis*. An inscription erected by *legio X Fretensis* in the Decapolis city of Scythopolis in 130 CE (§49). The deity Neptune, god of the sea, is depicted on the left. The inscription reads: Imp(eratori) Caes(ari) Traiano Hadriano Aug(usto) p(atri) p(atriae) leg(io) X Fret(ensis) coh(ors) I ("To the Imperator Caesar Traianus Hadrianus, father of the country. Dedicated by the first cohort of legio X Fretensis").

Photo credit: Charles Simon Clermont-Ganneau, used by permission.

Figure 3.2. Pig and Warship on a Brick Stamp of *legio X Fretensis*. A brick found at Giv'at Ram, a satellite village of Jerusalem. *Legio X Fretensis* produced these tiles 70–132 CE and included the image of a boar and war galley alongside an abbreviation of its name.

Photo credit: Photo credit: Yoav Dothan, used by permission.

Jesus' exorcism of Legion manifests a desire to expel the occupying Roman legion from the territory; the demon and thus Rome are invasive forces whose removal returns the man to health. Jesus' exorcism might read as an anticipation of God's kingdom, a kingdom in which Roman imperialism would have no part. The Dead Sea Scrolls also depict Roman military forces as demonic; the War Scroll depicts Roman soldiers as "the sons of darkness" under the command of the dark-angel Belial. This parallel in contemporaneous Jewish literature contributes plausibility to the military reading of the Gerasene Demoniac.

Though these two arguments may appear a strong enough basis to sustain the anti-occupation reading, there are important objections. Most notable is the complete absence of evidence for a *legionary* garrison near the city of Gerasa—where the story occurs—until 119 CE (§40; cf. §33, §41, §61), and therefore well after Mark's composition. Fretensis largely remained in the environs of Jerusalem after the Jewish War, though a single inscription attests a single cohort's presence in Scythopolis much later (§49; Figure 3.1). Indeed, there is no evidence that *legio X Fretensis* was ever in Gerasa; detachments of *legio VI Ferrata* and *V Macedonica* are attested, though these are also well after Mark's composition. Neither *VI Ferrata* nor *V Macedonia* used the aquatic and porcine symbols of *X Fretensis*, complicating their relevance for an anti-occupation subtext. Rather, the unit that was attested in Gerasa during the postwar period was *ala I Thracum Augusta*, an auxiliary cavalry unit (§§5–8). The narrative context of Gerasa is therefore difficult to link with Mark's reference to Legion, since there were no legions as such garrisoned there before or during the time of Mark's composition. The Gerasene military force was entirely another type, for which "Legion" was a misnomer.

There are also numerical discrepancies: the roughly 2,000 pigs have no obvious connection to military units in the region. Legions were nominally 6,000 soldiers, but the functional number was closer to 5,400. Standard (i.e., quingenary) auxiliary *cohortes* and *alae* were 600 and 480 men strong, respectively; double-sized (i.e., milliary) *cohortes* were 800 at full strength and *alae* were slightly smaller at 720 men. John Donahue and Daniel Harrington thus understand Legion merely as a numerical reference to the "many" demons in the man, as evident in the author's explanatory note (*hoti polloi esmen*; 5:9).[6] Donahue and Harrington prematurely dismiss the significance of the military entirely, but they helpfully identify the problems of Mark's enumeration.

Numerological objections do not entirely refute the anti-occupation reading, though. Some scholars contend that the author of Mark is unlikely to have known much about the different military units or their sizes, as he was not a military historian; similar mistakes are evident in Mark: we will see below that Mark strangely refers to a Herodian executioner by a Latin office, a matter that is unlikely historically, but may reflect Mark's context decades after the event. One should separate two distinct methodological questions: the military-historical problem of the army units in the region is entirely distinct from the sociology of knowledge question regarding who was likely to understand such distinctions. In the same way, say, an Iraqi civilian might characterize all American soldiers as "Marines," regardless of their military branch, so also might a refugee of the Jewish War characterize all foreign Roman forces as "legion," regardless of their composition and size. Writing from Capernaum, Mark was not in the immediate vicinity of Gerasa, so this sort of terminological slippage might be expected for someone writing about less familiar territory. Whatever the problems with Mark's depiction of a legion of 2,000 in Gerasa, it is possible that someone in his situation would make such a mistake. The inconsistencies between the military-historical evidence and Mark's account do not inspire confidence about the anti-military-oppression interpretation, but are not sufficient to discredit it entirely.

While military-historical factors do not present insurmountable difficulties for anti-occupation readings of the Gerasene Demoniac, much more significant problems lie in the narrative itself. Seyoon Kim observes that the anti-occupation reading only accounts for the first half of the pericope, up to Jesus' exorcism of Legion.[7] This reading has difficulty accounting for the Gerasene locals' rejection of Jesus after the exorcism (5:17). This final part of the pericope is conspicuously absent from most anti-occupation readings of the pericope, since within a military reading this would imply that Jesus' anti-Roman politics were an unwanted imposition. This problem is significant, since those rejecting Jesus' symbolic liberation from Rome are not elites, but apparently common denizens of the Decapolis. Richard Horsley is among the few to anticipate the ending's problems for his reading:

> The revelation that behind the mystification of demon-possession lay the Roman military as the real agent of their possession, however, was frightening to the community. They desperately begged Jesus to leave. It was difficult, indeed impossible, to face the real political-economic situation of imperial violence. Even though the hearers of Mark's story were hearing in this episode and others the "gospel" of God's liberation from Roman rule, they too would likely have felt uneasy and ambivalent about facing the concrete political-military forces that controlled their lives.[8]

In Horsley's framework, the city of Gerasa was subject to Roman oppression, and rather than come in direct conflict with the Roman Empire, a single man acted as a repository for the city's resentment of the Romans and adopted the subjectivity of a demoniac. Jesus exorcized the demon by naming the oppressive experience for what it is (i.e., Roman military occupation).

Horsley's explanation is difficult to accept, not least because of his unargued assertions—most important of which is that Jesus "reveals" the true identity of the demon as Legion, a revelation nowhere evident in the pericope. Horsley's reading also fails to account for why the denizens of the Decapolis eventually welcome the preaching of the former demoniac (5:20), despite their rejection of Jesus. Their reaction suggests this hesitation related more to the person of Jesus and his role in the pigs' death than the political implications of his exorcism. Furthermore, Horsley assumes the Gerasene residents' familiarity with twentieth-century anthropological insights and scholarly work on the link between psychosomatic illnesses and cognitive dissonance among colonized people: Gerasenes were unhappy with this information because it threatened "their delicately balanced adjustment to the Roman order."[9] The identity of the demon is troubling information *only if* the Gerasenes are consciously operating on the system theorized by Horsley. Horsley's explanation bounces between Mark's literary world and the real-life Gerasenes with whom the historical Jesus may have interacted;[10] this methodological confusion diminishes confidence in this reading of the pericope's conclusion.

Finally, this reading often elides the province of Judaea (and its prevalent Judaism) with the largely Gentile Decapolis. The two regions had very different postures toward the Roman Empire. Gerasa was a city of the Decapolis, a network of free cities north and east of Judaea. The Decapolis comprised several cities whose relationship with Rome was far more positive than it was with Judaea. The cities were mostly independent from their foundation under the Ptolemies and Seleucids, but were conquered by Alexander Jannaeus around 80 BCE and annexed to Hasmonaean Judaea. For some time, Judaea (not Rome!) was the imperial power operating in the Decapolis and the Hasmonaean program of Judaization was not well received in its cities. When the Roman general Pompey conquered the eastern Mediterranean in 64–63 BCE, he separated the cities of the Decapolis from Judaea and authorized their independence. There is every indication that residents of the Decapolis were thankful for this emancipation—not only do historical sources, inscriptions, and coinage attest such gratitude, but the cities of the Decapolis even adopted a Pompeian calendar in appreciation: most of the Decapolis took Pompey's conquest as their epochal year and enumerated their calendar beginning then.[11] Thus, the Roman army was apparently "greeted as liberators" in the Decapolis, as residents of the Decapolis experienced Judaea as a colonial power. The author of Mark lived near the Decapolis (e.g., one-day walk to Scythopolis, two-days to Gerasa) and was presumably aware of the very friendly attitude toward Rome in that region, an attitude evidenced at Mark's time in their staunch loyalty to Rome during the Jewish War. Mark's repeated insistence on the miracle's location near Gerasa—a city known for its Roman affinities—is thus difficult to reconcile with the politics of an anti-Roman reading. Thus, while Horsley and others are likely correct in suggesting the military plays a role in the pericope, scholars tend to simultaneously overdescribe and underhistoricize the military. The rejection of Jesus by the Gerasenes does not cohere with the anti-occupation reading of the pericope and attempts to explain it are unsatisfying. Also problematic is the failure to adequately address the pericope's location of Gerasa, as the Decapolis was known for its favorable stance toward the Roman Empire.

What can be said about the military in this passage, if an anti-occupation interpretation is not satisfactory? What is clear is that the military is given a privileged position that cannot be named directly, but is encoded within a different kind of discourse, one that is abstracted: Mark couches its polemic in a narrative about supernatural beings. Mark nevertheless offers this narrative as a topic of dispute. As William Arnal and Russell McCutcheon observe of other pericopae, "the effect [of doubt and inquiry in the Gospels] is to cue the hearer of such tales to their tentative, non-ordinary nature, and thus to stimulate further examination and exploration."[12] As examples elsewhere in Mark, Arnal and McCutcheon cite how Jesus' claim to forgive sins is met with charges of blasphemy by Jewish peers (2:7), which served to prompt further discussion among Markan Christians on the significance and authority of this particular practice. Likewise, Peter's bumbling confession at Caesarea Philippi (8:27–29) shows Jesus' identity to be an object of inquiry that might be addressed in different ways, some of which are preferable to others.

Mark may have sought to provoke further discussion about the Gerasene Demoniac. The primary controversy in the narrative is when the Gerasenes ask Jesus to leave because of "what happened to the demoniac and the pigs" (5:16), which inspired great fear. This might seem a logical starting point, but the ex-demoniac's preaching about "how much Jesus had done for him" (5:20) seems nearly synonymous, even though this message's reception was unambiguously positive. How might one resolve the tension between the rejection of Jesus and the preaching of the former demoniac? The reference to pigs is absent in the ex-demoniac's preaching, despite its presence in the explanation for Jesus' expulsion. The discrepancies between the reception of Jesus and the former demoniac may indicate that the loss of 2,000 pigs prompted the hostility to Jesus—that is, the destruction of the swine. This partially explains why Jesus was asked to leave the territory: though the exorcism may have eased travel for residents of the Decapolis, Jesus' action nevertheless resulted in loss of stock, notably the death of 2,000 pigs, and so interrupted other parts of life. Returning to Arnal and McCutcheon, the exorcism may have acted to encourage discussion of collateral damage for Markan ritual practices and the implications of such damage for outsiders. There is no evidence that Mark intended to provoke thought on the military (i.e., the Gerasenes' reaction has no obvious connection to "what happened to the demoniac and the pigs") and so military matters presumably occupy a secondary or tertiary position within the pericope. This is problematic for those who advance an anti-occupation reading; in such interpretations, Jesus' exorcism doubles as an exorcism of the Roman army and this connection between the military and the demonic is an essential (if not *the* essential) component of the story. This is difficult to sustain because Mark provides no indication that the military was a topic of controversy or even disagreement in the pericope. However the army figures into narrative of the Gerasene demoniac, the exorcism is not *about* the military.

How, then, might the military figure into the pericope? One rarely considered possibility is that the story alludes to the Roman army, but does so in a nonpolemical manner. While the word "legion" is a dead metaphor today that acts as a generic large numerical designation, this was not the case in postwar Palestine. To the contrary, *legio* seems to have first come to common use in Palestine after the Jewish War, as the word gained political salience at that point: Mark is the oldest known Greek Jewish text to use *legio* and the term has not been found in Latin Palestinian literature, inscriptions, and papyri until the Jewish War, either. One can thus assume that Mark understood the word in relation to the Roman army.

Mark is not alone in using military language to refer to the demonic. Unholy forces are described militarily in other texts of the Second Temple period, such as the *kittim* of the Dead Sea Scrolls' War Scroll and the watchers of 1 Enoch. The impulse to read these as inherently anti-occupation must be tempered with the observation that *holy* supernatural beings are often described similarly in early Christian and Second Temple Jewish literature. The reference to the twelve legions of angels at Jesus' command (Matt 26:53) and the heavenly host praising God (Luke 2:13–15) are hardly exceptional, since angels were regularly described in such terms by Second Temple Jewish authors. Not only do Michael and other angels perform military functions in numerous texts (e.g., War Scroll; 1 En.; Sib. Or. 2.214–237; LAE 40; 3 Bar. 4:7), but military terminology is regularly used to describe the hierarchical organization of angels.[13] For instance, God commands the angels to "gather before him, each according to his rank" in LAE 38.2, Philo assumes angels are organized into military ranks (*Conf.* 34.174), and the same can be said of various texts from Qumran (e.g., 4Q405, 4Q503). Arnal and McCutcheon suggest that "the fading presence of God in the early gospel literature must simply reflect the widespread distancing of deity in the Hellenistic and Roman periods . . . corresponding to the imperial distancing of centers of power and governance."[14] One witnesses an increase in the activity of divine delegates and functionaries during this period—angels, demons, and sons of God in Jewish literature—to mediate for an increasingly distant deity. As Arnal and McCutcheon imply, biblical conceptions of divine action are mediated by the author's experiences with terrestrial authorities, so the theological exchange of deities for lesser agents was partially shaped by similar shifts in administration-by-proxy under Greek and Roman Empires. Thus, while early Jewish literature imagined the Lord directly participating in battle against Israel's enemies (e.g., Josh 10, Ps 18:8–16), late Second Temple Jewish literature largely exchanges the Lord's personal involvement for his functionaries' activity (e.g., 2 Macc 15:22–23, Dan 10:10–13). To be sure, angels acted as the Lord's soldiers in earlier Jewish literature, but the shift of emphasis is clear and mirrors the changing role of heads-of-state with respect to warfare in the Near East. It is entirely predictable that the operation of the demonic would be understood similarly.

This schema of otherworldly proxy and delegation sees fruition in the Gerasene Demoniac. The pericope understands Roman legions as the most proximate functionaries of governmental power, and in turn represent the most proximate functionaries of supernatural power. This reading must be nonpolemical, since the Lord had his own legion of angels, as implied in Mark 8:38 and 13:24–27; the former imagines an imperial procession from in the heavens and the latter implies numerous angels at the son of man's command. Mark made use of the language available to him to describe both demons and the son of man's angels, doing so via military imagery. The interpretation proposed here, to tweak that of Pheme Perkins, suggests that the passage compares the demons to the Roman army, not vice versa.[15] The present interpretation avoids the problem with garrison placement that comes with anti-military-oppression interpretations; Mark's word *legio* need not relate to *legio*

X Fretensis specifically, but probably treats legions in general as a type of power-by-proxy. Thus, as elsewhere in Mark, ordinary people never encounter Satan—the ruler of demons (3:22)—directly, but only his functionaries.

This interpretation requires us to attribute the pig and aquatic imagery to coincidence. Misgav Har-Peled has written thoroughly on Jewish polemic against Rome that drew upon porcine imagery. After examining rabbinic and patristic evidence, he determines that

> the Midrashim and the Talmudim do not even mention *legio X Fretensis*, its boar emblems, the statue of a sow at the Jaffa Gate, or Aeneas' sow. If the boar emblem was "an insult thrown in the face of the Jewish nation," . . . we do not find any evidence of this in Jewish sources. The silence of the rabbinic sources makes it difficult to argue that the identification of Rome with the pig is a direct reaction to Rome's porcine symbols.[16]

If later Jewish writers, writing well into the garrison of Jerusalem/Aelia Capitolina by *legio X Fretensis*, make no allusions to Rome's porcine symbols in their polemic, it is all the more improbable that Mark contains an oblique reference to it, given how much more time the rabbis had to contemplate these symbols and the destruction of Jerusalem.

The demon's name Legion obviously evokes the military, which some take as evidence of anti-Roman polemic. There are other features that might be prima facie interpreted as relevant to the army, including the reference to the pigs and the aquatic setting of the story—which arguably evoke *legio X Fretensis*. But it becomes clear upon closer analysis that the parallels to *legio X Fretensis* and supposed criticism of Roman occupation are problematic. But rather than negating the significance of the military for the narrative, its import can be understood as occurring at a subtler level. This subtle mode of interpretation attends to the way in which the civilians' experience of the military was internalized and became a part of their conception of related matters—power-by-proxy and cosmic dualism, in this case. Thus, rather than a face-to-face encounter between God and Satan, Jesus engages Satan's demonic subordinates taking the form of Legion.

The Execution of John the Baptist—Mark 6:14–29

Two military figures appear during the beheading of John the Baptist: senior officers (*chiliarchoi*; 6:21) and an executioner acting on Antipas' behalf (*spekoulatōr*; 6:27–28). The senior officers are termed chiliarchs, which had long been used for military commanders—nominally of a thousand men. Roman literature and epigraphs use the term *chiliarchos* as a Greek translation of the Roman office tribune, but it is not clear if the term is Mark's or Antipas'. The matter verges on irrelevance because these officers are anonymous and functionless, simply acting as an indicator of the luxurious quality of Antipas' birthday party. The tribunes are pleased by Herodias' dancing and Antipas' desire to entertain his guests is one factor in leading him to execute John (6:26). One infers that such officers were not part of the author's experience of the military, as they represent a concentration of power that exists at a distance from exemplary characters (socially, temporally, etc.).

The execution presents a more complex set of issues. Mark transliterates the Latin *speculator*, which was most likely a term of Mark's own and not Antipas'; the pericope's historical accuracy is a secondary concern to Mark and it is unlikely Antipas' army was sufficiently Romanized to include this particular office. The duties Antipas assigns the *speculator* were not typical of army scouts, though ancient literature sometimes describes *speculatores* executing civilians.[17] Though the executioner is faceless, he has a clear function, beheading John, bringing his head on a platter, and giving it to Herodias. The executioner is accorded no narrative significance: Antipas alone is credited with the execution, with soldiers otherwise unmentioned (6:16), even when they act as an extension of Herod's power (6:17). The military is simultaneously both integral to the narrative, in that it provides Antipas his prestige and performs the act of execution, and irrelevant in the soldiers' inability to do anything on their own accord.

Much of the pericope is messy. The reference to Philip's mother-in-law and Herodias is confused and Mark implies the execution occurred in Galilee instead of Peraea (contrast Josephus *A.J.* 18.109–115, 136–137). More important to Mark's depiction than the historical proceedings of John's execution are broad polemics of characterization: Antipas as God-fearing but weak-willed, Herodias as temptress, and her mother as manipulative. Military representatives by contrast are of no special significance and occupy Mark's attention only insofar as they further other narrative goals: when Mark alludes to John's death later (9:12–13), the act of execution goes unmentioned, merely bieng a component of the generally poor treatment the new Elijah received. Soldiers are clearly not aligned with God's coming kingdom but neither are they especially aligned with Satan, acting with neither zeal nor hesitation.

The Olivet Discourse—Mark 13:1–37

Mark 13:1–2 contains the Gospel's most direct reference to the fall of the Jerusalem Temple. Though Jesus speaks to one of his disciples, informing him in the second-person singular that he will see the destruction of the temple (*blepeis*), there is no interest in who will carry out the destruction. John Kloppenborg has argued that Mark shows knowledge of *evocatio deorum*—a ritual wherein the Roman army beckoned out a tutelary deity before destroying a city or temple.[18] *Evocatio deorum* sought to dissolve a deity's relationship with the city he or she protected and welcomed that god into the Roman pantheon, assuring them worship among the Roman people. Around the time of the Jewish War, various people and texts claim that the Lord had abandoned Jerusalem (e.g., Luke 13:35, Josephus *J.W.* 5.367, 5.371, 5.412, 6.127, 6.348; cf. 6.300–309), suggesting that Titus performed the ritual before his siege of the Jewish temple. Kloppenborg contends that the most important part of the ritual was the utter destruction of the temple itself, such that no stone would be standing on another—something to which Mark 13:2 alludes.

Despite all of this, there is no indication that the Roman army is the object of blame or resentment in Mark. Mark 13:1–2 uses the passive voice to describe the fallen state of the temple, minimizing the agency of the men responsible for its destruction. Rather, Jerusalem brought this fate upon itself.

Mark 13:7–8 mentions a series of conflicts as portents preceding the son of man's coming. Once more, the actors are irrelevant for Mark's interests. But while it may be tempting to dismiss this passage as implying the reader was safe from these wars—after all, the reader is only hearing about wars and rumors thereof—we see a few verses later that this was not the case. The passage frames the war in a foretelling retrospect: the readers need not be alarmed about the war and other conflicts on account of their own well-being, as they have already survived it. The portents merely point toward the eventual arrival of God's kingdom soon after. The reader's proximity to the war is assumed and warranted no elaboration.

Also related to the war is Mark 13:14, which refers to "the abomination of desolation" (*to bdelygma tēs erēmōseōs*) that will stand where it should not. The abomination probably relates to Titus' siege of the temple in 70 CE, since Titus is the only person understood as a desecrator in other postwar texts (e.g., Sib. Or. 5.399, 2 Esd 11:1). While the significance of the temple's fall reverberates throughout the Gospel, Mark nevertheless identifies its significance principally as a portent preceding the son of man's coming, secondarily as an act of extreme blasphemy, and tertiarily as cause to flee Jerusalem. That is, as W. A. Such and others argue, actions of Titus and his army are important in that they precede the events leading to the son of man's arrival: "the Jewish War up to Titus' ruin of the city is preparatory for the trigger event, which 'minimizes' or 'clears the way' so that events up to Jerusalem's fall are subsumed under the impact of one event."[19] For this reason, Mark does not name Titus directly, instead adopting the phrasing of Dan 9:27 (LXX-OG and LXX-TH) and 1 Macc 1:54–56. Mark frames Titus' actions in terms of a predetermined course of events over which the Markan reader has no influence, but a prophecy that they have an opportunity to act upon. Titus' deeds thus have a cosmic significance, but are devoid of any personality; the siege of Jerusalem brings about a singular trauma, but the significance lies less in the act itself than in the effects felt throughout Palestine and the eschatological of which it is part.

The Jewish War receives further attention in the Olivet Discourse, though it continues to be imagined only in relevance to its victims' suffering: those in Judaea, pregnant women, the implications if during winter, etc. Brian Incigneri observes that "It would be a mistake to read 13:14b–18 as a lament over the loss of the Temple. Rather, it evokes *pity* for those caught up in the disaster."[20] The military is understood by contemporary scholars foremost as a war-machine, but it is striking that the Gospel of Mark—in its most direct references to the Jewish War—has no particular interest in implicating the army or Rome more broadly for the devastation of the conflict, nor even naming them as the reason for flight from Judaea. The Jewish War is repeatedly described in agentless terms, thereby complicating the common supposition that Palestinian denizens understood the military's primary role to be warfare.

Jesus and Pilate—Mark 15:1–15

The military is also active, if only implicitly, in the scenes with Pontius Pilate.[21] Similar to the execution of John the Baptist, the military's commander is responsible for the workings of his soldiers: once Jesus is handed over to Pilate (15:2), Pilate releases Barabbas, flogs Jesus, and hands him over to be crucified (15:15). All of these actions were presumably carried out by the Judaean auxiliaries, but Mark ascribes their agency to Pilate himself. The first two of these are not particularly surprising in that Jesus is taken into Pilate's custody, and Barabbas and Jesus are eventually released from it. The flogging is unusual in that Pilate's responsibility extends beyond the order to carry out the punishment to the act itself. Soldiers are entirely perfunctory in their role.[22] Pilate is a fundamentally ambivalent figure: on the one hand he is responsible for Jesus' punishment, despite his attempts to release Jesus from custody and awareness of the chief priests' machinations. If anything, Mark's ambivalent depiction of Pilate tends toward the positive, a man caught up in a controversy outside of his control, who must succumb to the will of the unruly masses to prevent further chaos.

Soldiers Mock and Crucify Jesus—Mark 15:16–32

Soldiers are also major participants in Jesus' crucifixion (15:21–32). They conscript Simon of Cyrene to carry the cross, bring Jesus to Golgotha, offer him wine with myrrh, crucify him, cast lots for his clothes, and crucify bandits nearby. The soldiers clearly mock Jesus in their offer of spoiled wine in 15:23. Aromatic wine was considered a delicacy at the time, so the offer of spoiled wine must be cruelty—probably with Ps 69:21 lurking in the narrative background. The soldiers' offer of bad wine to Jesus continues the parodic exaltation of the king of the Jews, as the soldiers had previously clothed Jesus' beaten body in a fine purple cloak. The soldiers' division of Jesus' clothes in 15:24 also draws upon biblical imagery, especially Ps 22:19—a Psalm especially significant for the Markan evangelist. The irony and unknowing acknowledgment of Jesus as true king continues with the placard of the charge against him (15:26), and while Mark is not explicit on who placed it, one may reasonably infer the soldiers did so.

Though the military is heavily implicated in the mockery of Jesus, Mark does not single out soldiers for criticism. Passersby, chief priests, scribes, and others being executed also ridicule Jesus about his alleged royalty. Moreover, the soldiers are never explicitly named as actors in the crucifixion scene: they are merely the implied actors in verbs with third-person plural subjects that require the reader's inference. Soldiers are not conspirators against Jesus and seem to have no special vendetta against him, but they nevertheless participate eagerly in his execution.

The Centurion's Confession—Mark 15:39, 44

Mark attributes Jesus' crucifixion to a centurion, a responsibility that is otherwise unattested for officers of that rank.[23] The centurion's acclamation of Jesus as God's son has attracted attention in Roman-oriented readings of Mark, as noted in the introduction. The passage has long been read as the first human acknowledgment of Jesus' divine sonship, the culmination of Mark's Gospel, wherein Jesus is finally and explicitly recognized as the son of God as prolonged misunderstanding; someone finally "gets it"! But many scholars now interpret the centurion's words as a sarcastic taunt to the effect of, "Some son of God this guy was!" Arguments supporting this more recent contention that the centurion mocked Jesus' death are numerous.[24] 1) Death by crucifixion was unlikely to persuade a bystander of divine sonship, as it would more likely indicate his insignificance to an impartial or hostile spectator (*idōn de . . . houtōs exepneusen*). Crucifixion was sufficiently humiliating that it would have been more likely to prompt mockery than inference of divine sonship, especially by an officer of the provincial administration that condemned him. 2) The confession appears in succession with other ironic scenes wherein the Jerusalem garrison unknowingly enacted a regal christology. The soldiers had already cloaked Jesus in royal garb, crowned him, hailed him as king of the Jews, and knelt in a sardonic act of obeisance (15:17–19). A taunt from the centurion could be another instance of dramatic irony in the passion narrative, as he unknowingly named Jesus' divine sonship for what it truly was. 3) The centurion displays an apathetic attitude toward Jesus later in the Gospel, when Pilate asks him whether Jesus is still alive (15:44–45). This apathy would be strange for someone who just proclaimed him as God's son. 4) It is intuitively unlikely that an outsider—one partially responsible for Jesus' execution—understood Jesus' christological significance, especially since those much closer to him were unable to grasp it. 5) The syntax of the centurion's claim indicates that Jesus *was* in the past God's son does not function well as a Christian confession (*alēthōs houtos ho anthrōpos huios theou ēn*; note the imperfect tense), assuming Mark understood Jesus' divine sonship to continue after death. Jesus was no longer God's son, according to the centurion. 6) A series of civilians taunt Jesus at his crucifixion: passersby (15:29–30), chief priests and scribes (15:31–32), and those crucified with him (15:32). The centurion may be yet another in this sequence. 7) It is not plausible that a Roman soldier, let alone a centurion, would violate expectations of emperor worship to acclaim another man God's son.

Some of these arguments helpfully moderate the prevailing reading of a confession-with-gusto. But while they may be instructive, they are not entirely convincing individually or as the basis for their conclusion. Indeed, proponents of the sarcastic reading often make hyperbolic or otherwise untenable claims to support their conclusion. The comments of Earl Johnson on point 7 above are worth quoting in this regard:

> Soldiers . . . took religious oaths to the Emperor, praising him as a god or a Son of God. . . . A Roman soldier's allegiance to the Emperor was expected to be absolute and it is unlikely that Mark's readers would find it believable that a professional soldier would risk his career in order to worship a crucified man, especially if by such a confession he might be risking his own death for treason.[25]

This claim is patently false, despite how frequently modern commentators assert variations on it. There is no evidence that if a soldier did not affirm the status of the emperor as *divi filius* or asserted another person held divine status he might be executed. As discussed in chapter 2, there was no expectation of exclusivity to the imperial or Olympian deities: numerous religions were represented in the military during the first century CE, including several that did not necessarily recognize the Roman pantheon, such as the cults of Isis and Serapis, Cybele, the Matronae Aufaniae, and the Jewish deity.[26] Because emperor divinity usually claimed lineage through the goddess Venus, cults which did not include her in their pantheon necessarily modified official Roman/Olympian claims, sometimes eschewing emperor divinity entirely.[27] There was no requirement for soldiers to practice emperor worship, despite the expectation of loyalty. This is worth repeating: Jews were predominantly monolatrous (i.e., worshipping only one deity) in the early Roman period to *the exclusion of emperor divinity*, yet Jewish soldiers and men in other non-Olympian cults are nevertheless attested in the Roman army.

The inaccuracy of this claim is evident in the examination of the army's approach to religious practices; if anything, the army was lenient with religious practices that did not egregiously violate soldiers' professional obligations. Oliver Stoll directs our attention to §51, an inscription dedicated to Deania Augusta erected by the legionaries of *V Macedonica* almost immediately upon their arrival in the city of Gerasa.[28] Stoll suggests this inscription to a foreign deity sought divine favor in an unfamiliar land; one could proffer similar interpretations of Lucius Obulnius' dedications to Zeus Kyrios and Atargatis Kyria (§14, §§17–18), as well as Diomedes' to Zeus Beelbaaros (§32) elsewhere in Palestine. One should also recall the ritual of *evocatio deorum* performed by Titus before destroying the Jerusalem Temple in 70 CE—the Jewish god was ritually "called out" prior to siege as an act of respect to the diety.[29] Examples of non-Roman religions in the army are widespread, with a sampling collected by Stoll. Stoll goes so far as to claim that the phenomenon of military dedications to local deities "can be observed in every military site throughout the empire." The combination of Roman tolerance with the difficulty of policing military religious practices meant that local deities were usually respected and sometimes revered among frontier garrisons. Earl Johnson's assertion that the acclamation of Christ as God's son (*huios theou*)—an appellation for a number of demigods and divinities throughout the empire—would have been regarded as subversive or treason is incorrect; indeed, Mark's knowledge of *evocatio deorum* indicates that the author was aware of the military's respect toward the Jewish deity.

There are other indicators that the centurion's confession was not mockery. Foremost is the fact that Mark consistently signals insincerity as such, usually explicitly. This is indicated through Jesus' successful riposte to a challenge (e.g., 2:5–12), overtly malicious behavior (e.g., 3:20–22), or the narrator's descriptions of opponents' intentions (e.g., 3:2–6). Literarily, there is no indication of malignant intent on the centurion's part, quite unlike Mark's depiction of the soldiers' beating of Jesus (15:20; *kai hote enepaixan autō*). Conversely, there are indications that Mark's centurion is to be understood as sincere. The most obvious indication is that the confession's statement followed a series of portents: Jesus' death cry, the darkness that accompanied Jesus' death, the ripping of the temple veil, and the speed of Jesus' demise.[30] Darkness at Jesus' death is particularly telling as a sign, since it was widely identified as a divine omen; indeed, Jesus' Jewish contemporary Philo claimed that eclipses were "indications either of the death of kings or of the destruction of cities."[31] Philo's understanding was common and evidenced throughout Greek and Roman texts, among which Jewish literature was no exception. Some combination of these four signs provides a more-than-sufficient literary basis to amaze the character of the centurion by way of Jesus' death. The succession of unusual occurrences at the time of Jesus' death is difficult to account for in a sarcastic reading of the passage: as humiliating as death by crucifixion may have been, it was surely overshadowed by the repeated occurrence of divine omens and the supernatural.

One should also dismiss the reading wherein the centurion's recognition of Jesus as God's son is incompatible with military service, leading him to disavow the imperial cult. Take, for instance, the remarks of Philip Bligh:

> Emperor Worship was obligatory on all (bar the Jew), especially on the soldier for whom it was the test of honour and obedience towards him who claimed his whole allegiance. . . . And it is left to a pagan soldier, a centurion, the backbone of the Roman army, from whom utter loyalty was demanded, who stands looking upward at the lacerated corpse of a Galilean peasant on a Roman gallows, to give the final verdict in the words of the imperial title: "*This* man, not Caesar, is the Son of God!"[32]

This is an inaccurate representation of military worship and religious practices for reasons noted above. Note also that emperor worship was never, at any point, mandated.[33] There are also literary problems with the idea that the centurion abandoned his post: the centurion continues to perform his duties as a Roman soldier by turning Jesus' cadaver over to Joseph of Arimathea (15:44–45). The continuation of his duties under Pilate indicates that Mark does not imagine military duties to be incompatible with the centurion's declaration, despite Bligh's assertion to the contrary. Likewise, Mark's centurion frames Jesus' divine sonship as a *past* state in the use of the imperfect tense within the confession, which is difficult to reconcile with the centurion's affirmation of Jesus' *continuing* significance that one finds in this interpretation.

But what is the significance of the centurion's confession if neither sarcastic denigration nor confession-with-gusto? A third, more tenable option is presented by Whitney Shiner, who understands the centurion's confession as a "near miss" of Markan Christianity.[34] The centurion adopts an outsider's respect toward Jesus, but he never approximates Markan Christianity; he "hears, but does not understand" (Mark 4:12; cf. Isa 6:9–10), like Pilate. Shiner notes several reasons to interpret the passage as something other than a full Christian confession, perhaps being a pagan quasi-acknowledgment of Jesus' significance. First, Shiner suggests that Mark was likely to understand the affirmation of Jesus as *a* God's son (*huios theou*)—with no definite articles—as a non-monotheistic affirmation of divine sonship that may evoke the Greek imperial title *huios theou*,[35] and thus christologically inadequate. Shiner's suspicions that *huios theou* signals a non-Jewish conception of God may be borne out in the Gerasene Demoniac's confession of Jesus as "son of the most high God" (*huios tou theou tou hypsistou*; 5:7). *theos hypsistos* was an epithet applied to the Jewish god, Zeus, among other preeminent gods and was not exclusive to Judaism.[36] Its use in the Septuagint and non-Jewish religions led "hypsistos [to be] used . . . as an appropriate name for God which could be put in the mouth of pagans in Jewish literature."[37] The exorcism of the Gerasene demoniac took place in Gentile territory and the phrase indicates the demoniac's non-Judaism. Mark therefore follows the Jewish literary practice of placing specific titles for deities in the mouths of Gentiles.[38]

Second, Shiner contends that the portents prompting the centurion's confession provided an insufficient basis for "authentic" Markan theology. That is, it was not enough to be amazed at the miraculous to be a Markan Christian, but required specifically Christian practices and beliefs—indeed, even Pilate was amazed at Jesus (15:6). Shiner's argument should be modified at this point. It may be more helpful to understand the centurion's confession as a misunderstanding provoked by a combination of Jesus' cry of dereliction *and* the portents at his death. The use of a past tense in the centurion's confession indicates that the centurion understood Jesus to no longer be (a) God's son, with the relationship presumably terminating at his death. If so, the centurion's christology corresponds with Jesus' quotation of Ps 22:1 and its lamentation of God's absence (15:34): God *was* with Jesus but has now forsaken and abandoned him; he was the son of God, but is no longer.

Some of those present at the crucifixion (*tines tōn parestētkotōn*; 15:35) misheard Jesus and understood him to be calling for Elijah. It is plausible that the centurion also misunderstood Jesus' cry in Aramaic, albeit differently from others present. Mark positions the centurion directly in front of Jesus (*parestēkōs ex enantias autou*), so the character could hear Jesus' dying words better than others nearby. The centurion understood Jesus' Aramaic, but did not recognize it as a quotation of Psalm 22. The centurion instead misinterpreted Jesus' cry as a straightforward description of divine abandonment. These factors are depicted as leading the centurion to assert that Jesus was a son of God, albeit in a past tense: up until the deity abandoned Jesus, he had been (a) God's son. As discussed in chapter 1, low-ranking auxiliaries in prewar Judaea spoke Aramaic (cf. Josephus *J.W.* 4.37–38). Mark knew that the Jerusalem garrison were auxiliaries during Jesus' life and it stands to reason that the author was also aware that Judaean soldiers understood Aramaic as well: Jesus spoke Aramaic to a man who was presumably Gentile elsewhere in Mark (i.e., *ephphatha* in 7:34 to a resident of the Decapolis).[39]

The centurion holds a distinctive position in the Gospel. Though he ultimately misunderstood Jesus' cry, unlike others who misinterpreted Jesus' last words, the centurion adopts a stance of respect toward Jesus and thus becomes a potential Christian. This potential, however, is squandered like the rich man in 10:17–22; he is nearly Christian, but not quite. The centurion's acknowledgment of Jesus' past sonship seems to hold no implications for his behavior, as his subsequent actions do not indicate a special attitude toward Jesus or his followers. But even so, the centurion is accorded greater personality and agency than any other military figure in Mark: even Titus' eventual destruction of Jerusalem was predetermined and initiates a sequence of cataclysmic events. The uniqueness of the centurion is consistent with the prevailing perception of the military among rural villagers in the Roman East, petitioning such men for aid and identifying them as distinct from the rabble comprising most of the army's ranks.

Conclusion

Mark was writing at a time when the impact of the Jewish War was still being felt and new social structures were starting to emerge in its wake. Mark's depiction of the military, however, is deeply ambivalent: most soldiers are part of a violent rabble that are hostile or at least dangerous, though the centurion is perceived as a man of potential openness and goodness—the only one without cruelty on his mind. Comparison with other texts, like Egyptian petitions to military officers and literary texts like Apuleius' *Metamorphoses*, indicate this was a common perception in frontier zones. Mark differs from elite literature in that high-ranking officers such as tribunes are not the "good ones" among the army's ranks, but instead are part of the faceless rich for whom the author has limited sympathy; in executing John the Baptist they are apathetic faces that populate Herod's party. The centurion, conversely, seems to be the sole person worthy of optimism, once again recalling the Egyptian petitions.

THE GOSPEL OF MATTHEW

The author of Matthew used two major sources in the composition of his gospel: the Gospel of Mark and the now-lost Sayings Gospel Q. Matthew wrote at a greater distance from the Jewish War than Mark, both geographically and chronologically, though he was likely Jewish like the authors of these sources. Matthew likely wrote from Syrian Antioch sometime between 85–95 CE and so his Gospel reflects a significantly different relationship with the military than either of his sources. We will see that whereas Mark depicts the military as a somewhat confused combination of prewar and postwar forces and reeling from the trauma of the Jewish War, Matthew has a somewhat more detached relationship with the armies of Palestine. As with Mark, scholars have recently tended to see Matthew as a text critical of the Roman imperial power; we will see the extent to which Matthew's reputation for anti-imperialism is earned.

The Slaughter of the Innocents—Matt 2:13–16

Herod the Great's slaughter of babies is not historically plausible. Though commentators often note Herod's reputation for cruelties upon his own people, including the assassination of three of his own sons and various other family members, it is unlikely that an event such as this would have been omitted from Josephus' historical record had it occurred. The explanation of R. T. France, that "terrible as such a slaughter might be for the local community, it is not on a scale to match the more spectacular assassinations recorded by Josephus," is inadequate.[40] Jews were known throughout the empire for their criticism of infanticide, an opinion that Josephus himself shares: "The [Mosaic] Law . . . forbids women to cause abortion of what is begotten, or to kill it afterward; and if any woman appears to have done so, she will be a murderer of her child, by killing a living creature and diminishing human kind."[41] The point here is not that it was impossible for a Jewish king to kill babies—there is evidence that Jewish soldiers, as with other armies of antiquity, performed infanticide on enemy populations (e.g., Ps 137:9). Rather, it is unlikely that if such a heinous act had occurred, that all memory of it would have been forgotten, aside from Matthew a century later—especially given how such a narrative would have contributed to Josephus' depiction of Herod.

But despite the fact that the story is unlikely to have any basis in history, Richard Horsley aptly summarizes an element of "realism" within Matthew's fantastical tale:

> The story of Matthew 2 comes to life vividly against the background of Herodian exploitation and tyranny. . . . He in fact responded with brutality to any threat to his own rule. . . . And Herod's response would have been cunning investigation and systematically efficient military action. . . . It is precisely against this background of exploitation and tyranny that the pre-Matthean and Matthean stories of the "massacre of the innocents" in reaction to the birth of the newborn king of the Jews originated and was cultivated.[42]

That is, despite the nonhistoricity of the Slaughter of the Innocents, this story is grounded in the memory of Herod as a client king who dealt brutally with any perceived threats to his rule of Judaea.

This memory also places Jesus as a competing "king of the Jews"—a title resembling that of the Herodian kings and princes. It is strange that the title ascribed to Jesus is most often an ethnic designation: he is king of the *Jews* (or Judaeans) and not king of a given territory. This peculiarity becomes clearer upon comparison with inscriptions of the Herodian dynasty: there is no evidence that they ever referred to themselves as king over the Jewish *ethnos*, even though Josephus twice refers to Herod the Great as "the king of the Jews" (*ho basileus tōn Ioudaiōn*).[43] Nor were the Herodian kings ever referred to as kings of Israel—a long-defunct kingdom. This is similar to "the king of Judaea" (*ho basileus tēs Ioudaias*), used by Herod the Great (40–4 BCE) and Agrippa I (41–44 CE), and recalls Archelaus' reign as "the ethnarch of Judaea" (*ho ethnarchēs tēs Ioudaias*; 4 BCE–6 CE). Thus, while Matthew's phrase "the king of the Jews" obviously relates to provincial politics, it does not directly appropriate any titles that the Herodians themselves used, at least as far as surviving data indicates. The extent to which these kings were attributed an ethnic title in the popular imagination is not entirely clear.

Matthew also uses the narrative to develop his intertextual theology that sees Jesus as fulfilling and reliving various episodes from the Septuagint. Most obviously, Matthew states that the slaughter of the innocents fulfilled the prophecy of Jer 31:15, evokes the baby Moses' escape from Pharaoh's infanticide, and places his father Joseph in the mold of Joseph from the book of Genesis. The narrative is a masterpiece of bringing together disparate political, theological, and historical interests into a single pericope.

Conscripted for a Mile—Matt 5:41

Jesus' saying about "going the extra mile" is among examples of turning the other cheek and not resisting the evil person (5:38–42). Conscription (*angareusei*) for a mile has long been interpreted as a reference to civilians being legally obligated to carry soldiers' supplies in the Roman provinces. Scholars commonly assume there was a law that mandated civilians carry a soldier's burden up to one mile, but there is no evidence for such laws and this inference is unnecessary. Laurena Ann Brink has demonstrated that such a meaning does not capture the legal sense of the word *angareia* in the Roman East during the first century CE, which differed from the common conception in important ways.[44] Namely, while officers could conscript others to carry their load, those conscripted were not civilians themselves, but civilians' animals; even then, they were only allowed to do so for official business—anything else was legally tantamount to robbery. Brink demonstrates that the supposition that civilians themselves were legally obligated to carry military goods is unfounded. She points to several inscriptions and papyri from the first century; one inscription erected 18–19 CE near Pisidia (*AE* 1976.653), for instance, clarifies that the number and species of livestock each rank were allowed: an equestrian officer was allowed three wagons or three mules, whereas a centurion was permitted one wagon or three mules. Regardless of the officer's rank, the edict indicates the payment due to the owner of the livestock, one sestertius per schoenus of mule travel (roughly 3.5 miles). Of course, just because anything beyond this was legally considered robbery does not mean it never happened. Ancient writers regularly depicted soldiers as bullies who demanded conscription from civilians they encountered (e.g., Apuleius *Metam.* 9.39) and the very existence of such an edict in Pisidia indicates that it was deemed necessary to prevent exploitation.

Matthew assumes that humans are the ones being conscripted to go a mile, not livestock. Perhaps someone is being extorted or illegally threatened to accompany an "evil person"—in this case, a soldier or officer. If so, this would imply a double humiliation should be accommodated: not only is the reader not being paid as would be legally required, but they themselves are being treated like livestock; it is easy to see why the person demanding such would be deemed "evil" by the evangelist!

Healing of the Centurion's Slave—Matt 8:5–13

One of the best examples of centurions being upheld for exemplary behavior is found in the Healing of the Centurion's Slave. Matthew adapted this story from the Sayings Gospel Q (cf. Luke 7:1–10).[45] The centurion is identified as a Gentile, who were probably few in prewar Galilee's royal army of Antipas, but the situation is not difficult to imagine. We have some evidence concerning the presence of centurions in the royal armies of Palestine (§13, §38). Because Capernaum was located on the border of Galilee and Batanaea, it is intuitively plausible that a centurion would be there to offer protection to customs agents and toll collectors, even if we do not have physical evidence of military presence until a century later (§315; cf. Table 2.1). Capernaum would be an ideal place for a regional centurion to be located: interregional roads crossed by the town, it was a port village on the Lake of Gennesaret, and it was on an international border; any single one of these qualities would warrant the presence of provincial administration and thus a centurion to ensure protection.

The non-Jewish ethnicity of this centurion is pivotal to the story: Jesus himself says the centurion exhibits more faith than anyone in Israel. Matthew is particularly invested in the centurion's non-Jewishness, with the inclusion of Matt 8:11–12, which in its Matthean context, condemns the presumptuousness of some Jewish groups and asserts that many Gentiles will enjoy the kingdom of heaven. The centurion's faith anticipates the apostolic mission to Gentiles (Matt 28:18–20), grounding it in the life of Jesus.

One strain of biblical scholarship proposes another interpretation of the pericope, suggesting it depicts the centurion in a same-sex relationship with the young man he asks to be healed.[46] At first glance, there appears to be little to justify a supposition of queer subtext: there is no explicit discussion of romance, sex, gender norms, or anything else of the sort. It may be helpful to walk through standard arguments for the queer interpretation regarding the pericope. The centurion's dialogue uses two distinct Greek words for "slave": *doulos* is employed in reference to slaves in general (8:9), but the word *pais* is found when the centurion refers to the slave boy who is ill (8:6; 8:8). The term *pais* not only referred to children and young slaves in Greek, but also to junior partners in male-male sexual relationships. This vocabulary in itself does not necessitate one to prefer the homoerotic sense of the word *pais*, but many commentators have observed the unusual concern that the centurion has for his slave. Finally, sex between men and slaves was prevalent in Hellenistic and Roman armies. While no one of these points requires contemporary readers to suppose the centurion was involved in same-sex intercourse, some interpreters contend that the whole is greater than the sum of its parts and so cumulatively indicate a sexual relationship between the centurion and his slave. The fact that Jesus says nothing about same-sex intercourse may imply his acceptance of queer sexual practices.

While not wishing to undermine queer politics, there are historical problems with this interpretation. Gospel scholarship commonly treats soldiers as though they were all legionaries—they are consistently depicted as Roman citizens, ethnically different from Palestinian civilians, relatively wealthy, etc. The problem with this depiction is the simple reason that legions were rarely in Palestine before the Jewish War. Typifying this problem is an article by Theodore Jennings and Tat-Siong Benny Liew arguing that Matt 8:5–13 suggests a same-sex relationship between a centurion and his slave, a relationship of which Jesus seems to approve.[47] I use this example not to argue one way or the other about Jennings and Liew's conclusion, but to examine the assumptions driving their argument. Jennings and Liew provide considerable literary evidence that legionaries had sexual intercourse with younger men (often slaves). The problem, however, is that the episode occurs in Capernaum, which was a part of Antipas' Galilean tetrarchy at the time and therefore garrisoned no Roman soldiers, but rather Antipas' royal forces comprising recruits from the local Galilean population.[48] Even though Galilee was a client kingdom that existed at the whims of Rome, the difference between Roman legionary and Galilean employment was not simply a matter of which administration signed the paychecks. Rather, the social systems that crucially structured the values of military institutions differed between client kingdoms and imperial provinces. A close reading of Jennings and Liew's primary sources reveals that the unequal relationship between Rome and the inhabitants of the regions occupied played a large role in generating the conditions for the use of slaves as sexual objects (cf. Figure 3.3). Denis Saddington observes that

Figure 3.3. Claudius and Britannia. Military conquest almost invariably resulted in sexual assault of the defeated in the Greco-Roman period. The emperor Claudius is here depicted as preparing to rape a woman—the personified Britannia—symbolic of his military victory in that region. The sculpture was erected in 48 CE in the Asian city of Aphrodisias (now in Turkey). A similar sculpture depicts the emperor Nero raping Armenia, again personified as a subjugated woman. Narrative depictions of such sexual assault in ancient literature are too numerous to list.
Photo credit: Steve Kershaw, used by permission.

"the instances they quote are of centurions raping adolescent boys in actual warfare or forcing themselves on unwilling young recruits." Polybius, for instance, tells of a centurion who captured a woman in Asia Minor, held her as a slave and repeatedly raped her—she eventually murdered him (21.38.1–6; cf. Livy 28.34). Hardly a relationship of mutual care.

The Roman imperial army was perhaps the single most transparent index of the emperor's might and thus served a significantly different purpose from the small army of a petty king; the norms of military masculinity into which the soldiers were socialized were therefore considerably different. This was especially with respect to how sexuality was understood in relation to maximizing the empire's maintainable dominion, preparing frontier regions for economic integration, minimizing violence from potentially hostile populations, affording of legal privileges to Roman citizens, etc.[49] Such sexual acts would be unlikely to occur in Herodian Palestine, since *none* of these components were constitutive of the military order—however salient they may have been for other apparatuses of Antipas' Galilean state. In fact, evidence suggests that the soldiers of the royal armies of Palestine found their sexual outlet in ways less apt to cause unrest among their fellow Galileans (let alone whatever punishment awaited an unruly soldier)—graffiti from soldiers garrisoned at Herodium indicate their use of sex workers and animal livestock for sexual gratification, the former of which is substantiated by Josephus.[50] It is more likely, then, that this slave is simply that: an unpaid laborer owned by a local military officer. Given the prevalence of slavery among the relatively wealthy of the Roman East, this would not be surprising.

Execution of John the Baptist—Matt 14:1–12

Matthew's execution of John the Baptist is a condensed version of what is found in Mark. Though much could be repeated from the discussion of Mark above, the Matthean version significantly reduces the role of the military in the story, omitting references to tribunes and the speculator who executes him. It is unlikely that these modifications are the result of any particular ideology or political program in Mark and simply the author's preference for a concise economy of words.

Olivet Discourse—Matt 24:1–51

Matthew's Olivet Discourse draws heavily upon Mark, revising and adding content, little of which concerns the military. Matthew's distance from the Jewish War is evident in how the evangelist relocates those sections most immediately relevant to the trauma of the war to a different section of his Gospel (e.g., Mark 13:9 to Matt 10:18), instead having Jesus claim the "sign of his coming" being related more to intra-Jewish conflicts—expulsion from synagogues, false prophets, preaching the Gospel to all nations/Gentiles, etc. Indeed, Matthew even revises the prompt for Jesus' discourse, as it is no longer *when* the temple will fall, but what will precede Jesus' return. For Matthew, the Jewish War itself was not a climactic event on the eschatological horizon, but one among several events that preceded Jesus' return. The evangelist's attention is instead focused upon the events befalling other Christians in his own day.

Swords at Gethsemane—Matt 26:52–54

Matthew's Passion Narrative builds upon Mark's, though it has some additions that are worthy of note. First, the question of Jesus and the sword, while present in Mark 14:47, is given much greater attention in Matthew. Since Jesus admonishes the use of swords at Gethsemane, it is in apparent tension with Matt 10:34–36, wherein Jesus proclaims that he does not bring peace, but the sword. Dale Martin notes that the interpretive inertia of Luke's narrative (I would suggest Matthew's as well) has led scholars to minimize the radical nature of Mark's Jesus.[51] Specifically, the Markan Jesus and his group had sufficient weapons among them that the narrator deems the entire group "armed." Matthew's redaction of the story includes a strong admonition against armed violence, which may imply that the armed disciple who harmed the high priest's slave is not representative of Jesus or his group—Luke pushes Jesus even farther in the direction of nonviolence, as will be seen below. Martin goes so far as to suggest that the arrest of (the historical) Jesus by armed legal enforcement is attributable to Jesus and his compatriots' possession of weapons, operating on the assumption that Rome disarmed conquered regions to prevent revolt. Martin's article generated a series of critical responses and, as an effort to explain why Jesus was executed both historically and in Mark, it is not entirely convincing.[52] At the literary level, Mark's depiction of Jesus and his associates as armed is curious, since the evangelist never explains why they are armed and it does not appear to be significant either before or after Gethsemane. It is thus difficult to read Mark as linking Jesus' weapons with his execution.

Matthew goes out of its way to distance Jesus and his followers from the endorsement of armed revolt. This verse is commonly linked with the fifth antithesis about turning the other cheek (5:38–42), though it has clear significance beyond that. The Matthean Jesus objects to the use of swords, not with a single reason, but with a series of unrelated arguments: 1) violence begets violence, 2) a more powerful means of overcoming our enemies is available, namely a divine army, and 3) the violence of this police force is necessary for the fulfillment of scripture. The subtext of violence is amplified by Matthew's use of the Latinism *legion*, putting in no uncertain terms the military capacity of God's angels. It is difficult to correlate these arguments into a single logic, as Jesus himself ultimately dies via a sword anyway—execution by crucifixion!—and there is no indication that this particular disciple dies by the sword despite his use of it.

The pericope distinguishes Jesus from a distinct-but-undefined group that uses arms (*pantes . . . hoi labontes machairan*; "all the sword-havers"). This group of armed people can be correlated with several different groups in Matthew's context of postwar Syria: the military, contemporary Jewish groups more open to the use of physical violence, retrospectively with freedom fighters in the Jewish War, or some combination thereof. Any of these would be consistent with the explicit linking of the death of Jesus with the punishment of Jerusalem (27:25), the framing of Barabbas as a "notorious prisoner," and clarifying that those who mock Jesus on the cross are also bandits (27:44). Though Matthew is deeply invested in divine violence, the same cannot be said of terrestrial violence, which consistently warrants condemnation—especially violence practiced by Jewish contemporaries.

A second interesting addition to Mark's Gethsemane scene is Matthew's reference to "twelve legions" of angels. Mark has some interest in spiritual warfare between divine and demonic beings—the son of man acts as the Lord's worldly delegate, under whose authority God has placed an angelic army (Mark 8:38, 13:24–27). John Kloppenborg thus concludes that "Mark represents Jesus as a military commander, defeating and dispatching demons, and as a strongman, able to overpower other men of violence."[53] Moreover, Satan rules over a demonic kingdom (3:22–27). This is even more explicit in Matthew, where Jesus uses directly military language to describe the organization of angelic groups. Though there were roughly thirty legions active at any given time in the Roman Empire with a comparable number of auxiliary soldiers as well, we need not suppose Matthew was privy to this knowledge—twelve legions is an impressive number and far outnumbers those present in Syria-Palestine.

Mockery of Jesus—Matt 27:27–31

Matthew's mockery of Jesus can be fruitfully explored via comparison with its Markan parallel (15:16–20), an episode significantly modified in Luke. There are two primary differences between Matthew's and Mark's narratives, both of which are more important than they initially appear: the color of the cloak in which the soldiers clothe Jesus and the extent of the abuse Jesus receives at their hands. In Mark, Jesus is robed in a purple cloak. The purple dye used in coloring the soldiers' cloak was produced in the eastern Mediterranean, mostly Tyre and its immediate vicinity.[54] The Phoenician coast was home to the mollusks *bolinus brandaris* and *hexaplex trunculus*, which secreted the fluid necessary for purple dye production. Purple dye was thus produced near Palestine, but it was prohibitively expensive—Pliny the Elder claims that a pound was priced at 100 denars during his youth, which skyrocketed to 1,000 denars per pound during his adulthood (*Nat.* 9.60–65). Pliny's prices are hyperbolic, but they may still be taken to indicate the dye's cost exceeded an auxiliary soldier's means.

Matthew changes the cloak's color to red, adding plausibility to Matthew's story in that red dye was more affordable and abundant throughout the empire, even if evidence of its military use in Judaea is lacking. Graham Sumner has collected the surviving evidence from early Roman Palestine, the sum of which indicates that soldiers usually wore white:[55] 1) §22 is a pay receipt that identifies linen tunics in use among soldiers at Masada, one of which is explicitly described as white. Graham Sumner notes that while linen was more expensive than wool, climatological concerns likely played a role in clothing decisions in Judaea. 2) Nearly all linen fragments excavated from Masada were undyed white, though it is not entirely clear which fragments were used for clothing and which were used for other purposes; it is also not clear which were used by military personnel and which were used by civilians. Regardless, one can assume some fragments are from military dress. 3) Josephus describes an incident when Simon bar Giora attempted to escape Roman capture during the siege of Jerusalem by impersonating a Roman soldier (*J. W.* 7.26–36)—Simon's imitation involved wearing a white tunic and a purple cloak (available to anyone wealthy enough to purchase one). 4) §189 is a receipt that includes the order of five white cloaks for Judaean soldiers. Unfortunately, no data precedes the Jewish War with any certainty. Mark depicts the soldiers clothing Jesus in a purple cloak, but this does not necessarily indicate the author was ignorant of soldierly garb in Palestine. Rather, one may take Mark's purple cloak as an indicator of military hubris and ironic acclamation of Jesus as king, as its color implies the wearer's wealth. Matthew depicts the robe as scarlet instead of purple, a decision that may reflect its Syrian provenance (or perhaps add to the blood imagery of Matthew's passion narrative). That is, the legionaries of Syria may have used red cloaks; if so, Matthew's redactional move here serves to add realism to Mark's depiction.

The mock robing of Jesus in Matthew and Mark might be illuminated by aristocratic polemic against soldiers' ostensive political aspirations. Elites of the early Roman empire commonly understood soldiers as tactless rabble whose rise in influence was a threat to the prevailing order, in that soldiers' and veterans' influence came at the expense of the old aristocracy. Elite authors attempted to discredit soldiers as a class by depicting them as an arrogant social type, marked by a desire for wealth and power, inflicting cruelty on any passersby—a stereotype that was taken up in popular writings as well. Mark's version of the pericope appears at first glance to be similar to aristocratic polemic. The use of the purple cloak creates an atmosphere of affluence, given the well-known expense of purple dye. The repeated acts of hubris during the beating were also familiar images to any reader of Apuleius or Josephus, among many other classical authors. There is also the irony of the soldiers hailing Jesus as king, oblivious to their acknowledgment of Jesus' true royal status. Despite the similarity to aristocratic polemic against soldiers, Mark's narrative is probably less connected with critique of soldiers' political ambitions than it is linking decadence with immorality (cf. Mark 10:17–31, 14:10–11). In a similar vein, high ranking officers acted as shorthand for the opulence of Antipas' birthday party in Mark 6:21. Mark does not appear particularly invested in the inherent immorality of soldiers, but he makes use of an existing stereotype for his own purposes.

The second way in which Matthew modifies Mark is by revising the sequence of events, such that mockery leads to physical violence. Immediately prior to this episode, Matthew added the note about Pontius Pilate washing his hands to absolve himself of responsibility, a character moment that contrasts the Judaean governor with his sadistic soldiers. Matthew further draws the reader's attention to the relationship between Pilate and his soldiers, noting that they were "the soldiers of the governor" (27:27). Matthew's juxtaposition of the saintly Pilate with the cruel soldiers also recalls aristocratic polemic against soldiers. But whereas Mark evinces little interest in the faults of soldiers as a class, the mere contrast between Pilate and his soldiers in Matthew generates such an effect. The fact that both centurions are positive figures in Matthew (8:5–13, 27:54) is sufficient to distinguish Matthew's depiction from elite literature, in that centurions were consistently depicted as equally corrupt as their subordinates in highbrow literature; that is, Matthew's dividing line between what constitutes the brutish underbelly of the Roman army and its goodly elites more closely resembles that of nonaristocratic literature, but it nevertheless operates on the distinctions salient to aristocratic discourse. Thus, whether or not Matthew *intends* to discredit soldiers by attributing them lowly vices in contrast with the nobility of the old elite, his depiction nevertheless *participates* in a popularized version of aristocratic polemic against hillbilly soldiers.

Matthew and Mark both adopt elements of standard polemic, but revise it for their own ends. Mark drew upon the imagery of a widespread critique of soldiers: soldiers' ownership of purple garments and their arrogant behavior toward others is too common a trope to attribute to coincidence. Mark seems to have little interest in attributing the soldiers' actions to a lower-class culture of violence, but readily employs imagery consistent therewith. Matthew contrasts the aristocratic innocence of Pilate with the sadism of his soldiers, even if he revises Mark's regal purple in favor of a less ostentatious red cloak.

Centurion's Confession and the Guards at the Tomb—Matt 27:54–28:15

The analysis of the centurion's confession in Mark above does not apply to Matthew. The Matthean centurion at the cross is explicitly motivated by *terror* of the surrounding events: earthquakes, supernatural darkness, resurrection of holy people, and the tearing of the temple curtain. Matthew appears less informed about the military situation of prewar Palestine and gives no indication that Gentiles might speak Hebrew or Aramaic; thus, it is unlikely he imagines the centurion might (mis)understand the last words of Jesus as the Markan did. Moreover, the confession in Matthew is not the centurion's alone, but includes others guarding Jesus at the cross. There is no indication that these soldiers or the centurion became Christians and the event has less narrative significance, as Matthew was less invested in the "Messianic secret" than Mark. Indeed, Matthew generally depicts soldiers as a failure-prone group of dimwits, a group for whom he holds little respect.

Matthew's negative characterization of common soldiers and neutral-to-positive depiction of Pilate continues after Jesus' death. At the request of Jewish leadership who hope to preempt tales of Jesus' resurrection, Pilate sends a contingent to guard the sealed tomb—the only instance where Pilate does so among the canonical Gospels. Matthew terms this group of soldiers *koustōdia*, a Greek transliteration of the Latin *custodia* meaning "guard," as these men are Judaean auxiliaries under Pilate's authority. Though Pilate is presented as a neutral arbiter, his soldiers fail at every possible instance: the soldiers "become like dead men" when an angel appears (unlike the women at the tomb, who are unfazed), they report to temple leadership instead of respecting the chain of command and addressing Pilate, and are easily bribed. Once again, Matthew depicts any soldiers below the office of centurion as untrustworthy rabble incapable of operating outside of their own self-interest, evident especially in the passion narrative. Much of the story stretches credulity (e.g., bribery in silver coinage, which was rare in prewar Judaea), but in a narrative sense, Matthew is able to accomplish much in this concise pericope designed to discredit rumors of a Christian conspiracy at the empty tomb.

Conclusion

The vast majority of Matthew's narratives about the military are taken over from its sources of Mark or the Sayings Gospel Q. Noteworthy is the omission of the name "Legion" in the story of the Gadarene Demoniac (8:28–34) and for this reason among others, it is difficult to assent to the idea that "Matthew uniformly unmasks Roman imperial power where he finds it."[56] Matthew instead evinces a fairly ambivalent stance toward the military. Though Matthew adds little of his own into Markan and Q passages, there is thoughtful redaction of existing stories, redaction that often points toward a disavowal of the revolutionary violence of the Jewish War. The depiction of the military and Roman officers is generally more positive than one finds in Mark and the condemnation of radical Jewish violence, not to mention low-ranking soldiers, is more explicit.

THE GOSPEL OF LUKE AND THE ACTS OF THE APOSTLES

Luke-Acts is a two-volume prose epic written by a single author, telling a story that stretches from the creation of Adam to the apostle Paul's imminent death. More than any other New Testament text, these books situate early Christianity within a Roman imperial context: the narrative is dated by events in Roman history (e.g., Luke 2:2, 3:1–2), characters refer to other noteworthy events happening throughout the Empire, early Christians regularly interact with imperial administrators, and the introductions to both Luke and Acts imitate the introductory prologues of Greco-Roman histories. Luke imagines a cozy relationship between Roman power and early Christianity, and scholars have consequently debated whether Luke-Acts is better characterized as *apologia pro imperio* or *apologia pro ecclesia*—whether the author is defending the empire's legitimacy for Christians or attempting to convince the empire that Christianity is not dangerous. Though this typology is reductive and can obscure the ambiguities and nuances of Luke-Acts' politics, it is nevertheless useful in its identification of Luke-Acts' overwhelmingly positive depiction of the Roman Empire and its agents.[57] The present section will not attempt to resolve this debate, but may illuminate how Luke-Acts' depiction of the military might further the discussion.

Luke-Acts was likely composed in the city of Ephesus 115–120 CE.[58] This date was late enough that the author could use extensive literary resources: Mark, Q, Paul's letters, Josephus, not to mention highbrow Greco-Roman literature (e.g., Vergil, Homer). Ephesus was a major imperial city, where a set of destructive Jewish revolts known as the War of Quietus were happening nearby in Mesopotamia, Cyprus, Syria, Egypt, Cyrenaica, and perhaps Judaea as well. This war comprised a series of local and largely uncoordinated Jewish revolts that lasted 115–117 CE. Our knowledge of these events is limited, thanks to spotty historical sources. This context may explain why Luke-Acts depicts most early Christians as Roman-friendly and Gentile-friendly Jews; whether to assure Romans that Christians had heritage among the "good (i.e., Roman-friendly) Jews" or to let Christians know that their founders had no connection to radical Jewish violence. Regardless, Luke is at pains to disparage anything resembling armed revolt.

Another set of issues we will encounter is the historical plausibility of the events Luke depicts. Second Temple Judaea expert Seth Schwartz claims that the author of Luke-Acts was "notoriously confused about the administration of Judaea."[59] The assessment of Roman military historian Jonathan Roth is more direct: "the historicity of Acts is a complex issue, but from a military historical perspective, the author of the work seems uninformed about the security situation in Jesus' Palestine."[60] Though neither Schwartz nor Roth are New Testament scholars, we will see below that their incredulity is warranted.

Repenting for John the Baptist—Luke 3:14

John the Baptist's call to repentance is met with questions from different groups as to how his message affects them: crowds, tax collectors, and soldiers. John's response to soldiers is almost comical by modern standards, "Neither extort nor defraud, and be satisfied with your wages." The remarkably low standard to which John holds soldiers warrants elaboration.

John's exhortation to not extort (*diaseisēte*) recalls the discussion of papyri in the previous chapter, as it was not uncommon for soldiers to do precisely that. *SB* 9207—an accounting ledger discussed in chapter 2—included entries for "extortion" (*hyper diaseismou*) at 2,200 drachma, with another at 400 drachmaes "to the soldier by demand." Likewise, an Egyptian scribe swears that a particular soldier is not extorting anyone (*diaseseismenō . . . hypo stratiōtou*; *P.Oxy.* 240), indicating this was a known phenomenon and this particular soldier was under suspicion. Many more examples could be cited. Extortion was most common in frontier regions, where the military could act with impunity, and in wartime situations, where the threat of soldiers' physical violence was palpable. In addition to extortion, wartime chaos provided ample opportunity for looting booty from temples and civilians; this was a practice of adding financial insult to physical injury. While this would not characterize prewar Palestine, which was relatively peaceful and garrisoned by locals already invested in the communities in which they served, it would certainly be relevant in

Luke's context, where the War of Quietus left considerable destruction for Jewish and Gentile civilians (see, e.g., §340)—a situation in which soldiers were apt to take advantage of the surrounding chaos for personal profit.

John's exhortation not to defraud (*sykophantēsēte*) uses a term generally associated with false accusations against the wealthy. Since soldiers saw a significant increase in status during the early imperial period, old-stock aristocrats felt threatened by the prospect of such a rowdy bunch gaining power. The Roman poet Juvenal expressed this anxiety well:

> Let us first consider the benefits common to all soldiers, of which not the least is this, that no civilian will dare to thrash you; if thrashed himself, he must hold his tongue, and not venture to exhibit to the praetor the teeth that have been knocked out, or the black and blue lumps upon his face, or the one eye left which the doctor holds out no hope of saving. If he seeks redress, he has appointed for him as judge a hob-nailed centurion with a row of jurors with brawny calves sitting before a big bench. (*Sat.* 16)

Juvenal depicts a kangaroo court where the perjury of a meathead soldier will find a sympathetic audience and the goodly aristocrats of yore have no chance of redress. These concerns were common in aristocratic literature of the early Roman period, though there is little historical evidence that it was a significant problem for such wealthy individuals.

Finally, wages were a matter of regular discontent among Roman soldiers. Chapter 1 argued that their pay was not particularly high and papyri indicate that soldiers were aware of pay disparities between auxiliaries and legionaries.[61] Moreover, auxiliaries were not eligible to receive *donativa*—bonus pay that emperors gave to legionaries either in appreciation for excellence in battle or as an attempt to curry loyalty in difficult times. The word that Luke uses, which is usually translated "wages," is an interesting choice, as it literally means "ration-pay" (*opsōnion*) and had a specific military meaning during the Hellenistic era (e.g., *P.Lond.* 23, *OGIS* 266, *P.Strassb.* 103, Polybius 6.39.12, 1 Macc 3:28) and occasionally in the Roman era (e.g., *BGU* 69, 1 Cor 9:7). Luke uses the term in a general sense of "military pay" rather than limited to food rations, evident in the triad of admonitions concerning generalized financial honesty that John gives the soldiers.

John's response to the soldiers' inquiry verges on ridiculous. As far as the Roman army was concerned, extortion, plunder, and loot were all authorized sources of income in the proper context. Over this section we will see that this is consistent with Luke's "utopian" depiction of the military. That is, soldiers and officers are open to hearing the good news from John, Jesus, Peter, and Paul and the military is a largely benevolent force that even educates Christians on occasion (e.g., Luke 7:1–10, Acts 10). This verse and others indicate that Luke is aware that soldier-civilian relations are not always positive in the "real world," but presents an imagined reality where these relations are rectified by the sheer persuasive force of preaching. However, it is not realistic to expect defining features of military conquest to be simply abandoned to accommodate the Christian life. In the sense that there is a "better way" of soldiers and civilians interacting, Luke-Acts can be deemed utopian in its imagination of social relations. That is, Luke-Acts is not simply an historical narrative depicting an idealized version of Christian beginnings; it does not merely use these figures in a normative way—that is, a realistic depiction of ought to be (i.e., we should do this or that)—but as a fantastical depiction of how things *can* be (i.e., what if this or that were possible?) when it comes to soldier-civilian interactions. This is evident from the removal of soldiers from the capture and imprisonment of John the Baptist only a few verses later (3:19–20), a move that alleviates any of their culpability in the capture and eventual death of a man who asked them to give up some of their worst (but most basic) habits. This is a consistent theme in both volumes of Luke-Acts.

Healing of the Centurion's Slave—Luke 7:1–10

The idealized depiction of soldier-civilian interactions continues in Luke's version of the Healing of the Centurion's Slave. The pericope derives from the Sayings Gospel Q and so has a parallel in Matthew. Much said about the Matthean parallel can be repeated here. One detail that is often mentioned that scholars is taken to support a queer reading of the story is the fact that the slave is described as "dear" to the centurion (*entimos*), though this additional word is hardly enough additional evidence to support such a reading.

More significant are Luke's substantial additions to the characterization of both the centurion and the Jewish leadership in Capernaum. The core of Luke's additions to the pericope are emphasizing the existing and remarkably positive relations between local Jews and the centurion: he is "well-regarded" and Jewish leadership beseeches Jesus on his behalf, "he is worthy, he loves our people and he built the synagogue for us." There is nothing implausible about a centurion being so well regarded among locals in prewar Palestine. We might recall from the earlier discussion how the people of the Palestinian city of Ascalon set up an inscription reading, "The council and people commemorate Aulus Instuleius Tenax, a centurion of *legio X Fretensis*, on account of his friendship" (§9) and an Aramaic graffito from a Jewish village adjacent to a military garrison reads, "May the memory of Lord Trajan be blessed . . ." (§75). Boaz Zissu and Avner Ecker note that the formula of this latter inscription was reserved for "esteemed persons in general and synagogue donors in particular."[62] The interactions that Luke depicts certainly did occur, but we will see are so consistent with Luke's politics of the military that one surmises these passages are a creation of the evangelist.[63]

Another point of difference from the Matthean version of the pericope is the significance of the centurion's ethnicity. Matthew concludes the story with a condemnation of Jewish rejection and the assertion that God's kingdom will welcome Gentiles instead. Luke is uninterested in anti-Jewish polemic here; the story instead anticipates another centurion, that of Cornelius in Acts 10, who is the first Gentile Christian. Both men have slaves, are publicly known for their respect for Judaism and hold the rank of centurion. At least at this point, the acceptance of Gentiles does not preclude Jewish participation in God's kingdom in Luke-Acts.

Gerasene Demoniac—Luke 8:26–39

Luke presents a more concise version of Mark's Gerasene Demoniac and the demon "Legion." The version in Luke has less political material than its Markan parallel, containing one reference to "Legion." The demoniac himself remains under guard, and the description of the demoniac is decidedly unmilitary: he is naked and homeless. Again, the pericope ends with Gerasene residents rejecting Jesus, but welcoming the preaching of the former demoniac. Luke's version, like Mark's, is difficult to read as a story of liberation from military occupation, especially since Luke was more distant from Palestine than Mark and depicts Roman soldiers in an almost uniformly positive manner. As was the case with Mark, it seems that the military metaphor(s) do not necessarily indicate a military subtext to the pericope.

Olivet Discourse—Luke 21:5–38

Like Mark and unlike Matthew, Luke's Olivet Discourse is preoccupied with the eschatological significance of the Jewish War and surrounding events. This is most evident in Luke's modifications to Mark's famously obscure "when you see the desolation of the abomination standing where it should not" reference (21:20). Here, Luke removes all of Mark's ambiguity that this refers to the Jewish War and the Jerusalem Temple's fall: "But when you see Jerusalem surrounded by military camps, then know that its desolation has come near." The destruction of Jerusalem itself becomes an eschatological event, albeit distinct from the consummation of history elaborated here as well.

The destruction of Jerusalem is explicit once again in 21:24: "[Judaeans] will fall by the edge of the sword and be led away as captives among all nations. Jerusalem will be trampled down by the Gentiles until the times of the Gentiles are fulfilled." The meaning of this final phrase is ambiguous: does "the times of the Gentiles" refer to the Roman rule that began during the war, Gentile supersession of the Jewish in Lukan salvation-history, some combination of the two, or something else entirely?[64] On a strictly literary level, it is difficult to avoid seeing a connection between the two uses of Gentile in a single sentence, which lends credence to the first interpretation. One might object that Jerusalem had been mostly under Gentile power since the banishment of Archelaus in 6 CE and almost entirely under Gentile power since the death of Agrippa I in 44 CE—the second volume of Luke-Acts indicates the author was acutely aware of this administrative situation. That is, if it had long been "the time of the Gentiles" in Judaea and thus Jerusalem and Luke knew this, then surely the phrase must refer to something else. This reasoning, however, does not consider the author's near-equivocation of "Jerusalem" with "the temple" throughout both

Luke and Acts. The reading of Roman domination may not describe the bureaucratic administration of Judaea/Jerusalem, but the demise of Jewish *religious* authority in the city of Jerusalem with the destruction of the temple and massive influx of Gentile administrators and soldiers that came about with the War's end. If this is so, it is among the few negative references to soldiers in Luke-Acts.

The alternate reading, wherein "the time of the Gentiles" refers to Gentile supersession of Jews in Lukan salvation-history is also tenable, despite how it necessitates Luke using the word "Gentile" both negatively and positively in the same sentence. Throughout Luke-Acts, Jews have an increasingly negative reaction to Christian preaching, Gentiles increasingly positive, and geographically Acts begins in Jerusalem and culminates in Paul's arrival in Rome: supersessionism is built into the narrative's ethnic and geographic arcs. These lend credence to a reading where the author expects the reader to suppose that Jews were totally eclipsed by Gentiles shortly after Acts' narrative conclusion: Luke-Acts' story ends around 62 CE and the temple fell eight years later. Luke's salvation-history may point this direction after the conclusion of the literary narrative—the speech of Stephen in Acts 7:2–53, for instance, locates God's glory outside the Jerusalem temple and recollects instances where the Jewish people were punished for mistreating God's messengers. The fall of the temple, for Acts, points the direction of Gentile Christians overtaking the Jewish people's place in the history of salvation.

Gethsemane—Luke 22:35–53

Luke, like Matthew, seems dismayed that the Markan Jesus and his disciples carried swords with them at Gethsemane. While Matthew addressed the issue by including an admonition against physical violence, Luke suggests their weapons fulfilled Isa 53:12. That is, Jesus' group *must* carry a sword—they need appear a legitimate physical threat in order to be crucified among "transgressors." Luke clarifies that neither Jesus nor those with him *normally* carried many weapons, as the two swords among the group are deemed sufficient. The evangelist implies that two swords were enough to prompt arrest without endorsing physical violence. This literary maneuver mitigates Mark's depiction of Jesus and his disciples as an armed gang. Thus, even when the sword comes out later in the narrative and the high priest's ear is lopped off, Jesus immediately heals the slave and admonishes the perpetrator of violence. Less easy to reconcile is the conclusion of the narrative, where Jesus expresses surprise at being arrested by armed men, which is taken over from Mark. It is difficult to avoid the impression that, based on Luke 22:35–38, the Lukan Jesus intended that exact outcome: arrest eventually leading to crucifixion; this may merely be an instance of what Mark Goodacre calls "redaction fatigue," a sort of continuity error resulting from heavy use of one source.

It is likely that Luke's saying about the swords in 22:35–38 has significance beyond explaining the loss of the slave's ear. Jesus began the saying by alluding to an earlier travel code that banned money, bag, belt, and footwear (10:4). Travel in

Greco-Roman antiquity was considerably more dangerous than in the current North Atlantic. Luke's earlier travel code advocated a remarkably confident way to travel, with only luggage sufficient for short distances with which one was greatly familiar (e.g., adjacent villages) or if one were willing to be harassed by passersby; without a bag to carry implements or a belt to carry any weapons for protection, Jesus indeed put them in a position of vulnerability by sending them out as "sheep among wolves" (10:3).[65] In short, the Lukan Jesus initially advocated an unrealistic way to travel beyond the shortest possible trips and this seems to have only applied to the period of his Galilean ministry. Luke 22:35–38 explicitly supersedes this earlier dress code in favor of a far more realistic one, a dress code that is normative for Luke and his readership. Rather than subjecting themselves to the whims of whatever bandits (or disgruntled soldiers!) might be lurking nearby, the Lukan Jesus permits basic self-defense for later Christians by demanding they carry a weapon. The saying thus has a dual purpose: an explanatory note for why the slave's ear was chopped off and a normative regulation permitting standard travel gear for Lukan Christians.

Finally, the presence of the "captains of the temple" at Gethsemane is peculiar. These figures are only present in Luke-Acts.[66] Despite the efforts of scholars to ascertain who these figures might be, Josephus, Philo, and other early sources indicate no knowledge of them within the temple hierarchy, and later rabbinic writings depict similar figures in a manner difficult to reconcile with the captains of the temple in Luke-Acts. Hints about their identity can be discerned through comparison with Luke's source Mark and Greek epigraphy. Luke's "captains of the temple" (22:52) replace Mark's "scribes" (14:43) in the list of those arresting Jesus. When one turns to Greek epigraphy of Luke's context—Asia Minor during the first centuries CE—one sees a similar title ("overseers of the temple") in inscriptions that refer to a panel responsible for the finances of a given temple. This is consistent with the depiction of the captains of the temple in Luke-Acts, as they are regularly involved in financial matters (Luke 22:4; Acts 4–5). Thus, while not exactly "scribes" in the expert-of-Jewish-law sense of the word, the duties of Anatolian overseers of the temple placed them in a scribal class of financial management.

Jesus and Pilate—Luke 23:1–25

Luke's amicable attitude toward Roman soldiers and administrators is fully evident in the trial scenes, where Luke goes to extraordinary lengths to both downplay Pilate's role in Jesus' execution and rehabilitate the image of soldiers. Luke begins the trial scene with a series of false accusations against Jesus: that he claimed to be a Messianic king, forbade tribute payment, and incited riots throughout Palestine. Like the Gospel of Matthew, Luke distances Jesus and early Christians from activities that might be perceived as subversive whenever possible. Luke accomplishes this by putting several additional false political charges in the mouth of his accusers, charges that Pilate dismisses as baseless. Luke uses this opportunity to relocate most of Mark's negative depiction of Rome at the feet of Antipas instead. To further exonerate Pilate (and thus Rome) from responsibility of Jesus' death, the Lukan Pilate sends him to Antipas, a move that stretches believability—indeed, the pericope never hints at any legal interest, with Antipas merely sending Jesus back to Pilate after humiliating him. Arguments that Antipas had jurisdictional authority in the situation are unconvincing, since offenders were tried in the province of their crime.[67] Moreover, the mockery that Mark had attributed to the Judaean auxiliaries is instead attributed to Antipas' royal soldiers—the dressing of Jesus in fine clothing, treating him with contempt, and mocking him verbally. These royal soldiers are presumably part of an escort for Herod's trip to Jerusalem for Passover and should probably be distinguished from the additional military presence generally found in Palestinian cities during festivals, attested in a number of sources (e.g., Josephus *J. W.* 2.224–227, *A.J.* 20.105–112, b.Šabb. 145b, Rab. [Lam] 1:52). The pericope intimates that Herod has a far closer relationship with his soldiers than Pilate had with the Judaean *auxilia*; that is, Antipas' men mock Jesus because of their commander-in-chief, whereas the auxiliaries do so independently of Pilate. Even later in Luke, when Jesus' clothes are divided and he is mocked, Luke does not explicitly make soldiers the agents of such activities in several instances, instead leaving an ambiguous third-person plural; "they" did these horrible things to Jesus, whoever "they" might be.

When Jesus is returned to Pilate, he refuses to execute Jesus, instead hoping to flog and eventually release Jesus—a desire he reiterates despite the protests of the crowds. Though the other Synoptic Gospels explain that it was custom to release a prisoner during the festival, Luke omits this. This has the effect of further alleviating responsibility from Pilate and transferring it to the crowd at Jerusalem. Throughout this sequence, Luke assures his readers that there is nothing "rebellious" about Jesus—Pilate explicitly says that he finds no basis for capital punishment and Barabbas is described as an insurrectionist guilty of murder, depicting him as a foil for the peaceful Jesus who only reluctantly brought two swords with him.

Soldiers and a Centurion at the Crucifixion—Luke 23:35–48

Though Luke largely frees soldiers and others linked directly with Rome (most especially Pilate) from culpability in Jesus' execution through subtle modification of Mark's Passion Narrative, this is not always the case. In one of the rare moments where Roman soldiers are depicted negatively, Luke describes them as mocking Jesus by offering him spoiled wine, though this is taken directly from Mark, combining the soldiers' mocking offer of wine and the gift of vinegary wine attributed to Jewish onlookers during his crucifixion. More interesting is the Lukan revision of the charge that because Jesus ostensibly claimed to be "the king of Jews," he should save himself, attributed to Jewish leadership in the Markan original, but shifted to the mouths of soldiers in Luke. This change is not merely the result of shifting subjects and verbs around, since Luke modifies Mark's phrasing, phrasing that was specifically Jewish in its orientation, to make it more relevant for the Judaean auxiliaries. No longer is the mockery that Jesus claims to be the "Messiah, the king of Israel" who might make his fellow Jews believe, but that he claimed a political title, "the king of Jews." Though we saw above that this is not quite the title that the Herodian kings took for themselves, it is close enough that it resonates with the political reality. The significance for the auxiliaries is obvious: were Jesus the king of Judaea, he would be their commander-in-chief and he could easily stop the execution with a single word.

Luke also modifies the centurion's confession. The centurion no longer proclaims Jesus "God's son" as in Mark, but merely "righteous" or "innocent" (*dikaios*). The significance of this confession is far more straightforward than Mark's ambiguous "truly this man was God's son," as Luke characterizes the centurion as "glorifying God." The phrase "son of God" is rare in Luke and Acts (though Jesus' divine sonship is unambiguous), as the evangelist prefers other christological titles for Jesus and generally depicts Jesus as a supremely righteous man (cf. Acts 3:14). This is especially so in the passion narrative, as several scholars have demonstrated that Luke—more than any other Gospel—depicts Jesus' execution within the schema of the "death of the righteous man," a trope common in Greco-Roman literature.[68] Laurena Brink makes a persuasive case that Luke depicts the centurion as a pious man (*eusebēs*); though obedient to his orders, the Gentile centurion nevertheless glorifies the Jewish deity.[69]

Cornelius the Centurion—Acts 10:1–48

Acts 10:1–48 depicts a "god-fearing" centurion named Cornelius in Caesarea who, in compliance with a divine vision, sent some of his soldiers to escort Peter from Joppa. Before the soldiers arrived, Peter also had a vision that revealed the arbitrariness of the categories "Jew" and "Gentile." Peter left with the soldiers and announced a correction to his previous teaching: Gentiles were no longer to be treated as unclean. The Cornelius pericope is one of the most important in Luke-Acts, since Peter's acceptance of Cornelius into the Christian community sets up Gentile fellowship as a legitimate enterprise in the early church. Acts' depiction of Cornelius echoes that of the centurion at Capernaum in Luke: they are centurions respected by local Jews for their piety, generous in their money to worthy causes, and humbly seek important Christians through intermediaries; the righteousness of Gentiles becomes apparent through each of them.

This pericope is among the most contentious in the New Testament among military historians. Most controversial is the question of whether there was an "Italian cohort" that garrisoned a centurion in Caesarea Maritima. The issue is complicated by a chronological factor—it is not obvious whether the incident occurs under king Agrippa I (41–44 CE) or Marullus the Roman prefect of Judaea (37–41 CE). The significance of the latter point is large enough on its own: if Agrippa I were king, then the army of Judaea would be Agrippa's royal soldiers, thus precluding the presence of auxiliaries from an "Italian cohort." But even if the incident occurred under Marullus, it is doubtful that there were any Italian auxiliaries in Judaea.

In an influential article, Michael P. Speidel argued that Acts' presents a plausible depiction of Judaean military units.[70] Acts describes Cornelius' unit as "the Italian cohort" (*hekatontarchēs ek speirēs tēs kaloumenēs Italikēs*; 10:1), revealing him to be an auxiliary. Speidel contends that this unit should be identified as *cohors II Italica c.R.*, which is attested in Syria around 63 CE, though it transferred to the province of Noricum before the Jewish War and returned to Syria after the Jewish War (§147, §§225–227, §§229–231). The citizen cohorts, of which this unit was one (indicated with the suffix *civium Romanorum*), were unique in that they were among the few auxiliary units comprising men who were citizens before their military service. That said, there was a reason they were auxiliaries and not legionaries: these auxiliaries were men of exceptionally low status (e.g., freed slaves, convicts). Against Speidel's claims, this unit's presence in Judaea contradicts the image presented by Josephus, who assumes that all soldiers in Judaea were Palestinian and mostly from Caesarea and Sebaste—including the royal army of Agrippa I, under whose reign this pericope probably occurs. Indeed, Josephus explicitly states that the Judaean soldiers before, during, and after Agrippa I's reign were Palestinian natives who had no desire to serve outside their homeland (*A.J.* 19.364–366). It is also hard to imagine why a cohort of Italian auxiliaries would have been present in a peaceful client kingdom with a sizeable army of its own. Regardless, the pericope reflects Luke-Acts' social agenda of Gentile integration so perfectly that it can be regarded as a creation of the author with reasonable confidence.

Luke depicts Cornelius and one of his subordinates as devout God-fearers. We saw in chapter 1 that there is ample evidence of Jewish soldiers in the Roman army and that the Roman military was tolerant of non-Olympian religions. Unfortunately, it is impossible to ascertain any evidence for Gentile soldiers who were sympathetic to Judaism in the surviving record, since military epitaphs rarely give any hint of a soldier's religious proclivities—evidence of Judaism is mostly onomastic: if a soldier's name is Aramaic or Hebrew (e.g., Simon, Bar-Simsus), we can surmise his ethnic background. Beyond this, one resorts to speculation: even funerary inscriptions erected for pre-Constantinian Christian soldiers begin with the standard pagan formula *dis manibus*: "to the gods of the underworld" (§344, §346). Indeed, pre-Constantinian Christian military inscriptions are often only discernable because of their provenience in a Christian cemetery—distinctively Christian phrasing or iconography are uncommon (e.g., references to God or the resurrection, images of crosses or fish).

Christians attitudes toward the military were mixed before Constantine. There is modest evidence of Christians in the Roman army beginning in the late second century, including a handful of inscriptions by or for Christian soldiers (§§342–362). This is not to mention the often-fantastical anecdotes in the writings of the church fathers (e.g., Tertullian *Apol.* 5, Eusebius *Eccl. hist.* 5.5). On the other hand, several church fathers condemned military employment as tantamount to idolatry; this is a recurring theme in the largely-fictional stories known as the Military Martyrs (e.g., *M.Marc.* 1.1, *M.Jul.* 1.4, *M.Das.*) and the polemic of the church father Tertullian (e.g., *Idol.* 17.2, *Cor.* 12). This denunciation of the army was not representative of all or most Christians, as the fathers represented a distinct

class of Christian intellectuals. The cumulative evidence suggests military employment was a controverted issue among Christians before Constantine's conversion,
since there is ample evidence that many Christians were soldiers at the same time
that their co-religionists were condemning such employment.[71] Indeed, the church
fathers roughly contemporaneous with Luke-Acts' composition neither condemn
nor praise military service when the topic arises (e.g., 1 Clem 37, 61; Ign. *Pol.*
6). It is not until near the turn of the third century that we find consistent and
unequivocal opposition to the military by Christian writers from the likes of Tertullian, Athenagoras, Origen, and Clement of Alexandria.

Soldiers of Agrippa I and Peter's Imprisonment—Acts 12:1–19

Scholars know far less about imprisonment during the early Roman period than
they usually acknowledge. *The Oxford History of the Prison*, for example, devotes a
single page to punitive detention during the early Roman Empire.[72] In lieu of any
definitive work in the late twentieth or early twenty-first centuries on imprisonment
in the early Roman Empire, some New Testament scholars have performed a service
by filling in gaps left by classicists on the topic.[73]

Greco-Roman prisons differed significantly from modern ones. First, there was
no concept of a "prison sentence" in antiquity, as these facilities merely acted to hold
the accused or convicted until their trial took place or sentence was carried out (e.g.,
beatings, crucifixion, banishment). The implications of this are enormous: there
was no distinction between jails and prisons, it was rare to be in a prison more than
a couple of days, conditions were nightmarish, torture was used to elicit confession,
etc. Second, imprisonment was often ad hoc outside of large cities, because there
was no building dedicated to criminal detention; consequently, any secure room
might serve that purpose if need be: a treasury, a small temple, or an abandoned
house might be reasonable candidates. Third, jail conditions bore little resemblance
to those of the modern North Atlantic, where prisoners' rights are an important,
if often disregarded, concept. Few jailers provided food, so one depended on family and friends. Violent offenders were put in the same room as others (sometimes
leading to assault, both physical and sexual), there was no recourse for anything that
happened while detained, and conditions were dismal in other respects. Fourth,
imprisonment was often a humiliating experience, where dignity and honor were
difficult to maintain. As we will see with Paul's letter to the Philippians, prison led
many to either carry out or contemplate suicide as a way of maintaining honor;
one thinks of Socrates voluntarily drinking the hemlock to kill himself rather than
escape prison. Finally, elite Roman citizens were generally exempt from prison stays,
instead permitted the much more comfortable experience of house arrest. Perhaps
this last point is not so different from our own context where "affluenza" is a feasible
defense in a criminal trial.

While Peter's experience of prison in Acts 12 is not pleasant, it is far better than the
experience of real-life noncitizens in prison: Peter interacts with no other prisoners, he

is able to leave (miraculously) in less than a day, he does not fear what might happen after his arrest, etc. Indeed, every instance of imprisonment in Acts is a far less painful experience than one would typically expect. Acts is less interested in a "realistic" depiction of jail than it is an edifying and entertaining narrative, as Acts' scenes more closely resemble ancient fiction than more realistic depictions of imprisonment that we find elsewhere. Saundra Schwartz offers a compelling analysis of the type-scenes common to both Greek novels and Acts' trial and prison scenes, including this pericope.[74] The scene takes place before Agrippa I, who is depicted as a violent despot that rules by whim; Peter is unjustly thrown in jail, yet miraculously escapes to freedom. Because Agrippa's royal soldiers fail to find Peter, the impulsive king executes them. Though Peter does not undergo a full trial, Agrippa offers a substitute for the prosecutor's speech found in the novels and is soon struck dead by an angel for hubris. This pericope draws upon various tropes found in fictional trial scenes, but is particularly similar to the fictional trial of Melite and Clitophon (Achilles Tatius 7.7–16): a capricious ruler, a plucky hero jailed on false charges, grand speeches to the assembly, etc. Though the novels consistently depict lovers being kept apart by the fickleness of *Fortuna*, this romantic relationship is transposed onto Christians and their love of Christ in Acts.

Agrippa's royal soldiers are entirely incidental in this passage.[75] They carry out the whims of Agrippa and when they fail to do so, their execution is further evidence of his petty and hubristic character. The depiction of Agrippa and his soldiers is fully Lukan, as Philo and Josephus characterize the king more positively. That is, the soldiers' role is more to showcase the cruel capriciousness of yet another Jewish king, than make a statement about the military. Unfortunately, evidence concerning early Roman prisons is sparse and inconsistent, with Judaean prisons being even more so. Consequently, it is difficult to ascertain the extent to which these soldiers carried out their duties.

Paul on Trial—Acts 21:17–22:30

Chaos ensues when Paul returns to Jerusalem from Ephesus, as his presence provokes Anatolian Jews shortly after his arrival. A tribune of the Judaean *auxilia* named Claudius Lysias intervenes to save Paul's life, but detains him in the barracks until the situation becomes clearer. The prewar Jerusalem barracks are well known, located at the northwest corner of the Jerusalem temple, known as the Antonia Fortress (see Figure 3.4). Lysias is predictably friendly, accommodating Paul and being measured in his actions, even permitting Paul to give an extended oration on the steps of Antonia to the incensed crowd. Only when the crowd proves unruly once more does Lysias order Paul flogged.

The narrative takes a surprising turn during the flogging scene, as Paul uses his Roman citizenship as a *deus ex machina* to escape punishment. Even setting aside the reasons for suspecting Paul's Roman citizenship was a Lukan invention,[76] the episode is peculiar in several ways. First, tribunes like Lysias, unlike nearly all auxiliaries, had Roman citizenship during their career. Indeed, they were not mere citizens, but were

Figure 3.4. Antonia Fortress. A model reconstruction of the Antonia Fortress in Jerusalem, the temple courtyard is visible to its left. This was the barracks of the Jerusalem garrison in the prewar period, built 31 BCE and named after Herod the Great's ally Mark Antony. It was destroyed during the Jewish War.
Photo credit: Malini Kaushik, used by permission.

usually part of the equestrian order, sometimes even senators. These officers were drawn from all over the empire, though it was common for Easterners to stay in that part of the empire, especially under the emperors Claudius and Hadrian.[77] This is a helpful coincidence, as this story occurs under Claudius and Acts was probably written in the early years of Hadrian's reign. Hubert Devijver observes that equestrian citizens usually became military commanders through either family connections (if one came from influential stock) or through friendly relations with influential individuals. Despite the claims of Acts in this incident, there is almost no evidence of purchasing Roman citizenship (22:28). The closest one comes is a humorous anecdote related by Suetonius about a Greek man named Staberius Eros. The passage is ambiguous, but most interpreters suppose that Eros sold himself into slavery to a Roman citizen, whom Eros had bribed to manumit and grant citizenship immediately afterward (*Gramm.* 13; cf. Petronius *Sat.* 57); bribery of administrators, of course, was another possibility (Cassius Dio 60.17.5–6). Given that the tribune is explicitly the first citizen in his family, it is doubtful Luke imagines he achieved his post via family prestige. Those benefiting from the friendship of even more influential men could advance via one of three career tracks:

A. The equestrian officer might perform the whole of his service in a single province, owing his promotions to the judgment of successive governors.
B. Governors, on transfer to other provinces, might take selected equestrian officers with them.
C. Such transfers might be made direct by the emperor, through his secretary *ab epsitulis*, to fill establishment vacancies with suitably qualified officers.[78]

Devijver notes that the first of these was most common in the first century; the cognomen Lysias is Greek or Syrian, indicating Luke imagined his character originated in the Eastern half of the empire. The Judaean governor at the time this story is depicted, Marcus Antonius Felix, had his first command in Judaea, rendering Devijver's second possibility moot—Luke's use of Josephus may have permitted knowledge of this biographical datum. This, of course, is working on the assumption that Luke had modest knowledge about the career structure of Roman officers, though one might reasonably dispute this. All of this has the effect of rendering Paul as a Roman citizen par excellence—he has deeper Roman roots than the military commanders working on behalf of the Roman government. What Paul may lack in the Lysias' wealth and power, he can claim in possession of Roman citizenship and knowledge of Roman law.

Second, in a move that is typical of Luke, pedestrian characters are eliminated from the story to focus on the appeal of Christianity to elites. In this case, the reader is left wondering who was supposed to flog Paul. We can infer that "they" are subordinates of the centurion standing nearby to witness the beating. This tendency is evident elsewhere in Acts: the Ethiopian Eunuch (Acts 8:26–40) may be an even more egregious instance of Lukan chauvinism: Philip ignores less prestigious characters (e.g., the narrator fails to even mention the chariot driver goes) to focus on the man who advises the Ethiopian queen. All of this indicates the author's desire to show a deeply amicable relationship between Christians and "legitimate" people in or proximate to power, even if it is at the expense of commoners.[79]

Finally, it is worth noting the continuing subtext about the Roman military's fear of Jewish potential for revolutionary/anti-Roman violence. When Lysias first meets Paul, he inquires whether he is the Egyptian prophet (21:38), a figure Josephus describes as a revolutionary who came to Jerusalem in order that he might persuade others to overthrow the Roman garrison (*J.W.* 2.261–263; *A.J.* 20.169–171); Lysias is also ready to appease the unruly crowds, though refusing to break Roman law in order to do so. We will see this continues in the following section, where an attempted lynch mob is thwarted by Lysias and others.

Paul's Escort to Caesarea—Acts 23:1–35

Most biographical information about the tribune Claudius Lysias comes from this section, wherein he directly commands his troops and sends a letter to the governor Felix in hopes of escorting Paul to safety. We can infer that Luke depicts him as the tribune of a *cohors equitatae*—a cohort that included both infantry and cavalry, as he sends two centuries of soldiers and a nearly two turmae of horsemen to bring Paul to Felix in Caesarea; this vexillation may also include 200 *dexiolaboi*, an otherwise-unknown word. Though there is no evidence of *cohors equitatae* in prewar Palestine, it was not uncommon for such units to simply drop the "equitatae" from their name; it is thus plausible but impossible to verify.

The force that Lysias devotes to Paul's escort is massive to the point of absurdity. At that time, *cohortes equitatae* consisted of six centuries (nominally 100 infantry each, but actually 80), and four turmae (40 cavalry each): Lysias sends nearly half of the troops under his command to ensure the safety of a single man! Regardless, Lyias' overzied escort for Paul successfully prevents a violent outburst. Some commentators object to Lysias' general behavior as a bungled plan that averted disaster not by Lysias' tactician mind, but through dumb luck. To take a single example, Hans-Josef Klauck objects to the honesty of Lysias' interactions with the Sanhedrin in strong terms:

> In his dealings with the Sanhedrin, Claudius Lysias by no means proceeded in such a deliberate and well-planned manner as it appears in [the letter written to Felix]. It is also almost a miracle that he gained the impression from the Sanhedrin's clamorous discussion in Hebrew that the debate concerned only inner-Jewish controversies about the interpretation of the law, but not the questions of public safety and order for which alone he was responsible as a Roman official.[80]

This interpretation relies on a *realistic* reading of the episode. That is, Klauck and others treat the episode as though it were bound by the conventions of reality as opposed to the conventions of literature. Klauck observes that it is unlikely that a tribune would know Hebrew well enough to ascertain the contents of the Sanhedrin's conversation, but if Luke tells the story not as history but as literature that serves to entertain and edify, then this and other problems are rendered moot, as they are extraneous to Luke's narrative goals. Likewise, the common objection that Lysias' letter overstates his role in preventing Paul's flogging or at least minimizes how Paul was nearly beaten unlawfully, does not understand the matter within the fundamentally positive characterization of Lysias. Acts has little room for the moral ambiguity of any named characters and there is no reason to think that Luke is characterizing Lysias as deceptive in this pericope. Rather, Paul's massive escort and rose-tinted letter should be understood as part of Luke's depiction of Lysias as not just a fundamentally good man, but a typically good *Roman*.

Julius the Centurion—Acts 27:1–28:31

When Paul is transferred from Caesarea to the city of Rome, he is escorted by a centurion named Julius of the Augustan cohort. Julius, unsurprisingly, is a very positive character in the story, protecting Paul when things go awry on the ship, even if it seems disproportionate to send a centurion to escort a single man across the empire. Archaeologists have recovered a number of inscriptions attesting a unit named *cohors Augusta* in Agrippa II's kingdom of Batanaea (§12, §14, §18, §30, §197; cf. §17, §201). This identification is lent credence literarily by Agrippa's significance in Paul's voyage.

Michael P. Speidel suggested that Rome allowed one of its auxiliary cohorts to operate under the authority of Agrippa II alongside his own royal army, pointing to the kingdoms of Bosporus and Palmyra.[81] As noted above, this was not a common phenomenon. Agrippa was an especially loyal client king—famously aiding Rome in the Jewish War—and it was unusual for Rome to send forces unless there were concerns about the king's authority or civic unrest, as with the reign of Archelaus (Josephus *A.J.* 17.299). However, the reverse seems more likely: that the army of Agrippa II aided Rome in one or another military venture (likely Nero's aborted Parthian campaign or the Jewish War) and so his soldiers were probably subsumed temporarily into the Roman *auxilia* under the aegis of *cohors Augusta Canathenorum*, though this is deeply uncertain. One of the soldiers in *cohors Augusta*, Diomedes son of Chares, seems to have been part of a Jewish cavalry that had long served in Batanaean client kingdoms (see Josephus, *Life* 35, 37).[82] This would explain why some inscriptions mention the more prestigious auxiliary cohort and some only mention Agrippa's army: only after being conscripted into Rome's army did Agrippa's soldiers operate under that name. If this suggestion is correct, then the *cohors Augusta* attested epigraphically came into existence in 55 CE at the earliest, but more probably in 67 CE during the Jewish War. There are significant reasons to doubt both of these scenarios, so caution on the matter is necessary.

Attempts to identify the historical military unit of which Julius was a part are also problematically rooted in "realistic" readings of Acts that do not adequately attend to the literary function of such details. The episode is full of rich, even heavy-handed, symbolism if treated as a literary construct: Acts begins in Jerusalem and culminates in the city of *Rome* where he will see the *emperor*; to get there, Paul is escorted from *Caesarea* by a centurion named *Julius* of the *Augustan* cohort. Analogously, imagine a military officer named George who serves on the USS *Constitution* escorting an ambitious preacher from the state of Washington to the District of Columbia. That is, the sheer density of allusive names and significant locations suggests something other than historical recollection is going on. Names in Acts are significant for characterization and symbolic import, as was convention in ancient literature.[83] For instance, the Acts Seminar suggests that the healing of Aeneas in Acts 9:32–35 employs similarly allegorical devices vis-à-vis the Christian gospel and the Roman Empire; Aeneas had been bedridden eight years (on the eve of Rome's 800th anniversary, it might be added!) until Peter healed him. Given that a far more famous Aeneas figured into legends about Rome's foundation, the implication is obvious: the Christian gospel will heal a long-ailing empire.[84] In the case of Julius, Rome is at its best when it protects Christians, even in seemingly dire circumstances.

Luke-Acts concludes with Paul's house arrest in the city of Rome, under guard of a single soldier—one infers this is not Julius, but a common foot soldier. Though soldiers guarded individuals under house arrest, house arrest was reserved for only the wealthiest Romans, ones who would not be properly thrown in a makeshift jail with common criminals, but deemed worthy of a degree of freedom, albeit under supervision. Though one does not get this impression from his letters, Acts' depiction of Paul is certainly one such man: a well-educated, law-abiding Roman citizen who is comfortable conversing with society's elite.[85] Even so, the entire episode is confusing; Richard Pervo jokes: "After eight chapters (Acts 21–28) focused on Paul's legal problems, the reader no longer understands why he is under arrest, of what he was really charged, why he appealed in the first place, or why he did not withdraw his appeal later."[86] Even in its finale, Acts defies a "realistic" reading.

Conclusion

Among the canonical New Testament, Luke-Acts has the most cohesive depiction of the military.[87] Though soldiers are not without flaw, they are overwhelmingly positive and exemplary figures. The evangelist reduces the soldiers' cruelty in the passion narrative, Jewish guilt (rather than Roman might) is responsible for the temple's fall, and the conversion of soldiers anticipates the spread of Christianity not only to Gentiles but to the Roman Empire as a whole. In the words of Alex Kyrychenko, "the Roman empire in Luke's narrative is a receptive mission field, and the Roman centurion, the principal representative of the Empire, exemplifies the desired response."[88] One might contrast this with the depiction of Herodian kings and their soldiers, who are uniformly malignant: Antipas executes John and behaves like a child when he meets Jesus and his soldiers (not Romans) are responsible for mocking Jesus; Agrippa I likewise executes James and attempts to kill Peter, while his soldiers are bumbling fools.[89] While one might interpret this as Luke moderating his own pro-Roman stance, it should probably instead be read as part of Luke's polemic against the idea that anyone other than Jesus could deserve the title "king of the Jews"; when contextualized alongside Luke's dismissal of other Jewish messianic claimants of the prewar period (e.g., the Egyptian [Acts 21:38], Theudas [Acts 5:33–39]; cf. the more Roman-friendly Baptist [Luke 3:14]). Indeed, these violent and capricious men act as foils for the sanguine and self-controlled figure of Jesus in Luke.

Furthermore, it is important to emphasize that while scholars may think of a unified "Roman army" or client kingdoms as functional extensions of the Roman Empire, this is not how ancient writers thought of the matter. This should not surprise us since, as Simon James demonstrates in a series of brilliant articles, the notion of a singular "Roman army" is a modern construct that is entirely anachronistic to the Roman Principate. James contends that the notion of "the Roman army" is problematic because it retrojects modern notions of the "war machine" into antiquity and purports a unified military apparatus that did not exist in the Roman Empire. James notes that there was not even a word that communicated the notion of a single, monolithic military in the early Common Era:

"Army" (*exercitus*), singular, was used for a particular grouping of forces, such as the standing army of a province or a corps specially assembled for a particular campaign. When generalising about the military, they employed plurals, writing of "the armies" (*exercitus*), "the legions" (*legiones*), "the regiments" (*numeri*), etc., and not least of "the soldiers" (*milites*), denoting a socio-political category.[90]

The same can be said of Greek. *stratia* and *stratopedon* were used in a manner similar to *exercitus* and generalizations about military units are consistently pluralized, such as *speirai*, *teloi*, *lochoi*, and *stratiōtai*, rather than operating with any singular and unified referent. This identification occurred not only at the linguistic level, but the social one as well: soldiers understood themselves to be clients of their specific general and individual generals likewise understood themselves to be patrons of their soldiers; the effects of these social relations were abundantly evident under the trium-virates and during the Year of the Four Emperors, not to mention soldiers claiming to be the son of their general in military diplomas (e.g., §295). It was simply not feasible for emperors to act as effective patrons over the entire army, thereby limit-ing any overarching unity among their ranks. Thus, one would not expect Luke to treat Herodian and Judaean provincial forces as part of a single military apparatus, however inextricable we might find the two today.

Whether the sum of Luke's political ideology is better characterized as Christi-anity accommodating Rome or the belief that Rome could accommodate Christi-anity is debated, but it is difficult to sustain a reading that sees them as the object of Lukan critique.

THE GOSPEL OF JOHN

The Gospel of John is difficult to discuss because of its complicated compositional process. John was likely composed in a series of stages, evident in the Gospel's deep investment in not only the miracles of Jesus, but the interpretation of those miracles and at times the interpretation of that interpretation. The precise nature of these stages is contested, though here I assume a three-part compositional history: 1) a prewar Galilean collection of five miracles; 2) a narrative Signs Gospel dependent upon at least some of the Synoptic Gospels (probably Palestine in the 90s); and 3) finally the canonical John with its "I am" sayings and extensive speeches, written somewhere in Syria in the first decades of the second century.[91] All three compo-sitional levels are preserved in the canonical Gospel of John, but are only really relevant for the Healing of the Royal Official's Son; the passion narratives mostly correspond to the second literary level and do not warrant a composition-historical commentary beyond the "canonical" level. As for the military, scholars have noted that there are no Gentiles in John until the passion narrative! Thus, it should come as little surprise that soldiers barely figure into narrative. Because John is less in-vested in this institution than the other canonical Gospels, it is difficult to ascertain the politics of the military therein.

Royal Official's Son—John 4:45–54

The Healing of the Royal Official's Son resembles the Healing of the Centurion's Slave found in Matthew 8:5–13 and Luke 7:1–10: an agent of the state who owns slaves and resides in Capernaum asks Jesus to save a young man's life; Jesus heals the boy from a distance, discussing the importance of belief in the process. There are important differences between the Johannine and Synoptic version of the story: the centurion is a Gentile military official desiring his son be healed, whereas John depicts a Jewish royal official (?) whose son is ill. Despite these deviations, they may attest a single tradition that precedes the canonical Gospels or the Sayings Gospel Q, perhaps even going back to the historical Jesus himself.

John's pericope has two distinct hermeneutical levels, corresponding to the compositional development of the Gospel described above.[92] The first stage consists of several miracle episodes absent a larger narrative framework; at this point, the miracle stories are not accompanied by the extensive speeches found in the Gospel of John, but short episodes that highlight the novelty of Jesus bringing together an unlikely group of Galileans. These stories are similar to, but distinct from, Mark's miracle chains; they contain three healings, feedings, and a water episode. But whereas in Mark these stories primarily concern the wonder of those nearby and Jesus' identity as God's son, in John such narratives concern group formation. Jesus and his followers are firmly placed within Judaism—there is no substantive antagonism against other Jews in this first phase. Scholars often refer to this as the Signs Gospel, though I would argue it is a misnomer at this compositional stage; at this point, the miracles were not yet deemed "signs" (*semeia*) and Jesus' significance was that of a community founder, not as God's unique son.

At this first hermeneutical level, the story concerns a "royal official" (*basilikos*) whose son is sick. The term "royal official" is sufficiently vague in Greek that it is difficult to ascertain what this might mean with much specificity. Josephus, who also lived in Galilee before the war, used the term frequently and with reference to different types of people—royal soldiers (e.g., *J.W.* 1.45, 2.429), members of the royal family (e.g., *J.W.* 1.249), and royal administrators (*A.J.* 16.399). John could be evoking any of these meanings, though scholars tend to prefer the last of these possibilities. Since the first stage of John was probably composed in prewar Galilee (likely Cana), one assumes the author had familiarity with the realia of Galilean villages. Capernaum was a border town for most of the prewar period and thus presumably at least a few administrators inhabited the town: high-level toll collectors, scribes, bureaucrats, and subregional governors might reasonably take residence there, though the lack of surviving sources renders it impossible to know with any certainty (see Table 2.1). This is complicated by the fact that Capernaum appears to have been a largely obscure village in antiquity: Josephus makes only incidental mention of it (*Life* 403) and it is otherwise unattested in texts independent from the Gospels until centuries later.[93]

The presence of these administrators and bureaucrats was reason enough to station at least a few royal soldiers, offering protection to toll collectors and other border agents at the Galilee-Batanaea border; the Sayings Gospel Q (also produced in prewar Galilee) depicts a military officer there as well. Thus, it is entirely plausible that John's *basilikos* is a soldier or military officer. We know of no legislation preventing royal soldiers from marrying or having children while in service—we know that auxiliaries often did so—so that element is certainly plausible. As with the Matthean/ Lukan parallel, it is entirely plausible that an officer in the royal Herodian army might own slaves, as it is well attested among Romans of comparable rank.[94]

The second stage of John's Gospel built upon the core five Galilean miracles, providing it a narrative framework (including a passion narrative) that drew upon the Synoptic Gospels and added two miracles. But despite the addition of a more robust biographical framework, the miracles remain pivotal. The miracles are no longer merely important for their role in founding a community, but the miracles become a sign of Jesus' significance in themselves. The import of these miracles is accentuated by their connection with Jewish holy days that Jesus metaphorically embodies (e.g., healing a blind man and the festival of lights, feeding and Passover). At its second level, the Healing of the Royal Official's Son is linked with the "festival" mentioned early in the pericope, which was earlier established to be Passover (2:13, 4:45). The theological significance is obvious: death passes by a Jewish child, not because of lamb's blood on the door of his house, but on account of belief in Jesus' word. The miracle becomes a "sign" of Jesus' power over life and death. Note also at this level the significance of a royal official humbling himself before Jesus: Jesus is called "Lord/Sir" and the official begs Jesus to heal his son: the word *kyrios* was merely a mode of respectful address at the first level, but is imbued with Christological significance at the second.

Gethsemane—John 18:1–12

The first Gentiles to appear in John are the soldiers present at Gethsemane. John draws upon the Synoptic Gospels for this episode, leading to substantial overlaps. John is unique among the canonical Gospels in placing the Judaean *auxilia* at the arrest of Jesus, as the Synoptics instead depict Jewish leadership arresting Jesus. The exact numbers are unclear, but John implies it is substantial: a (whole?) cohort, along with its tribune—not to mention Pharisaic officers and the priestly group. The reason for the soldiers' inclusion is not obvious, as it adds little to the proceedings, except to display Jesus' power: when Jesus utters the divine name, they and others at Gethsemane fall—apparently involuntarily—into a position of supplication (18:6). The implication seems to be that the soldiers and Jewish leadership that accompanies them are aligned with demonic forces overwhelmed by mere invocation of the Lord's name. Jesus submits to the soldiers and Jewish leadership because they perform an integral role in the divine plan that culminates in his death and resurrection (18:11). The soldiers and their tribune say nothing, acting as part of a faceless group, and seem to be throwaway references that merely add to the narrative's ominous tone.

The episode with the swords and slave's ear is also present in John, though it takes on typically Johannine significance. Jesus rebukes Peter for lopping off the slave's ear (named Malchus in John) not because of its violence, but because Jesus accepts the cup that the Father offers. Unlike Matthew and Luke, where the episode either implicitly or explicitly condemns practices of armed violence, John is more invested in how it reveals Jesus' place in the ensuing cosmic drama.

Pilate, His Soldiers, and the "King"—John 18:28–19:22

Pilate is a more interesting character in John than the other Gospels. He has extensive dialogue and is ascribed greater theological significance than the Synoptic accounts upon which John bases his story. Central to his character is the issue of the two kingdoms: Jesus' heavenly kingdom and how others misunderstand this to be an earthly kingdom. Throughout this section, Jesus' "kingdom" is placed in contrast with earthly politics; this is especially evident when Pilate must choose between being a friend of the emperor (*philos tou Kaisaros*) or implicitly being a friend of God and Jesus. It would seem that one cannot have the proverbial best of both worlds, as these are mutually exclusive realms of power in John's dualistic worldview. Jesus is repeatedly referred to as "king" (*basileus*) and there is an ironic regal aspect to the proceedings, even more so than in the Synoptics: the cross becomes his throne, he is clothed in a purple robe and hailed king of the Jews by soldiers, Jesus exudes an otherworldly and "true" power, the chief priests declare the emperor their king, Pilate is unable to change the epigraph from "king of the Jews" to a less provocative charge, etc. Pilate is unable to comprehend the distinction between earthly and heavenly kingdoms, even though this was a common part of Greco-Roman discourse in which Jupiter/Zeus was referred to as the king of the gods. Though heavenly and earthly authority was fairly conventional in ancient literature, Pilate does not understand it. It does not seem that John intends to depict Pilate as dense, instead using Pilate's reaction to highlight Jesus' otherworldly wisdom.

The charge that anyone claiming to be king also sets himself against the emperor, a charge linked with Pilate's status as a "friend of Caesar," is significant. Implicit within this is the idea that Jesus' kingship is legally tantamount to anti-Roman sedition and that the Jewish crowd rejects Jesus' claims of kingship as a sign of loyalty to Caesar. The former is confused on two points: first, emperors authorized many client kings and, secondly, emperors actively distanced themselves from the title of "king." On the first point, the title of "friend of Caesar" (*philokaisar*) was often given to client kings who kept themselves in Rome's good graces, including the Jewish kings of the Herodian dynasty, who displayed this epithet on their inscriptions and coins.[95] John may have in mind the revolutionaries of the Jewish War who claimed kingship within Palestine against Rome, as indicated in their wording: those who *make themselves* king (i.e., apart from the emperor's personal authorization) work against Rome. John, however, is more invested in the second of these meanings. More than any other canonical Gospel, John depicts non-Christian Jews as not only opposed to Jesus and early Christians, but actively evil and satanic in their mission (e.g., 6:64–71, 8:44, 13:2).

There are likewise problems with John's association of the title king (*basileus*) with the emperor. The Roman Principate (27 BCE-284 CE) was defined by the fact that emperors actively distanced themselves from monarchic ambitions, emphasizing ostensive continuity with the Roman Republic. It was only with the rise of the Dominate (284 CE) that emperors abandoned the pretense of preserving the Roman Republic, but even then the emperors did not use the title "king" themselves. Despite widespread assumptions that the Greek word *basileus* was readily applied to the emperor, Albert Wifstrand showed it was an aberration in Greek texts of the first century, used in highbrow literature of the second century, and only in common use by the third century.[96] It was not until the Heraclius I (610 CE) of the Byzantine Empire that emperors adopted *basileus* as an official title; the Latin *rex* was never used in this capacity. Monarchy was antithetical to Roman values during the first three centuries of the Common Era and Jewish writers almost always respected this nuance.[97] It is thus peculiar that the chief priests hail the emperor as their king. Presumably John uses the term in a loose sense for narrative and christological purposes.

Soldiers at the Cross—John 19:23–37

John's passion narrative depends on the Synoptic accounts, though adding and revising several elements. Especially important to John's narrative is the soldiers' role in inadvertently fulfilling scripture. The soldiers divide and gamble for Jesus' clothes (19:23–25), which the evangelist explicitly connects with Ps 22:19; they pierce Jesus' side after his death (19:37), which fulfills Zech 12:10; and, finally, they do not break his bones (19:33–37), fulfilling Ex 12:46 and Ps 34:20. This is not to mention more subtle evocations of the Hebrew Bible. Soldiers are unwitting participants in a divine drama; they have no real agency of their own, simply doing what God's will requires of them.

That said, John deviates from his Synoptic sources at times. John derives his division of garments from his Synoptic sources, but adds that it was divided into four— one for each soldier; the purpose here seems simply to add a level of realism to the scene, though this need not be taken as historical. Michelle Christian discusses the tendency of Greco-Roman and Christian authors to use numbers to create a degree of verisimilitude, being produced in "a culture where everything of importance was quantifiable."[98] There is no obvious literary or military reason for the number four here (four soldiers is half of a single *contubernium*, the smallest unit in the Roman army), so it is most plausibly interpreted as the evangelist's attempt to render the narrative credible.

Luke and John share the detail that soldiers offer Jesus spoiled wine, which is attributed to a Jewish bystander in Mark 15:23 and Matthew 27:34. Unlike the Synoptic accounts, John does not depict this as ironical, mocking, or especially cruel: Jesus explicitly says he is thirsty and they provide him drink. This is soaked in a sponge and placed on a branch of hyssop. Raymond Brown notes that the mention of hyssop is confusing, since this plant is not known in Palestine, nor is there reason to think a branch of the relatively small bush could support the weight of a wine-soaked sponge.[99] There have been various attempts to resolve this peculiarity and particularly influential is a biblical manuscript from the eleventh century (476*) that renders John 19:29 as *hyssos* (spear) instead of *hyssōpos* (hyssop). Many commentators and some translations (e.g., New English Bible) have accepted this variant and emended the text of John thus, despite significant reasons to doubt its originality. As G. D. Kilpatrick observes, the Greek word *hyssos* corresponds to the Latin *pilus*, a spear that was used by legionaries and not auxiliaries that were present in prewar Judaea.[100] Though it is conceivable John included this as an anachronism, it does little to resolve the problems of historical plausibility. However, Bruce Metzger makes an important observation that the origin of this particular textual variant is accidental and not an attempt to resolve the peculiarity of John's narrative here, a haplographic error in particular—the omission of repeated letters in a manuscript.[101] The use of hyssop, frankly, is confusing and difficult to explain satisfactorily.

Another interesting addition to the Synoptic passion narrative is the refusal to break Jesus' legs. Commentators often note that a crucified person might have their legs broken to speed up their asphyxiation—no longer able to push up with their feet to alleviate pressure on their chest to breathe. Evidence for this practice, however, is entirely post-Johannine, and most of it depends on John itself (e.g., Gos. Peter 4.14, Acts Andr. 54, Aurelius Victor *Caes.* 41.4). Indeed, the discovery of the bones of a crucified man, evidently with broken legs, was initially taken as support for the practice, though this is now disputed; some archaeologists think that the victim's leg bones were broken postmortem. As Brian Johnson has demonstrated, the vast majority of scholars depend on the late and Christian author Lactantius (*Inst.* 4.26) for their interpretation of John.[102] When most ancient authors discuss the practice of *crurifragium*—the breaking of leg bones as a part of execution—they do not have crucifixion specifically in mind. Polybius (1.80.13) describes men who were executed by having their legs broken, hands cut off, and were thrown alive in a trench, unable to escape; the plays of Plautus depict slaves who had their legs broken to prevent escape from capital punishment (*Poen.* 886, *Truc.* 638). All of this suggests that *crurifragium*, regardless of the means of execution, was a penalty reserved for the socially marginal: slaves, pirates, and probably those deemed worthy of crucifixion. The spread of the practice as well as the means of its enactment during the early Roman period, however, remains uncertain.

Finally, John adds the postmortem stabbing of Jesus' corpse to release blood and water. The practice of piercing the crucified to verify their death is attested in a pseudonymous text in the name of the first-century Roman rhetorician Quintillian. Pseudo-Quintillian narrates the story of a father who wanted to bury the corpse of his son who washed ashore, against the will of the boy's mother. The father argues, "As for those who die on the cross, the executioner does not forbid the burying of those who have been pierced, whereas pirates did not more than throw the corpse into the sea" (*Decl.* 6.9). The date of pseudo-Quintillian's text is uncertain, but it attests to the practice found in John. The church father Origen seems aware of it as well:

> Pilate . . . fearing the tumult of the entire mob, did not command, according to the Romans' custom for those who are crucified, that Jesus be pierced underneath the arms of his body—which those who do occasionally sentence individuals that have been found guilty of serious crimes (since accordingly they who are not pierced after being nailed endure great agony, but live on in extreme pain, sometimes indeed an entire night and even a whole day after). (*Comm. ser. Matt.* 140)

Though Origen knows the biblical accounts, he articulates knowledge of this practice that may be independent of John, lending modest credence to its plausibility.

Conclusion

Soldiers in John are a means to an end, literarily speaking; as Gentiles, they hold little significance for the author. The Father asks that the Son be crucified in order that scripture might be fulfilled. Romans, entirely oblivious to their role in the cosmic (or even scriptural) narrative, do their duty. Soldiers mostly appear in material adapted from the Synoptic Gospels and seem to be worthy of little attention, as they—along with their commander Pilate—are largely exonerated for their role in the execution of Jesus. Instead, John prefers to belabor the conflict between Jesus and his Jewish peers, likely reflecting tension between the author and Jewish people in his own social setting.

NOTES

1. Christopher B. Zeichmann, "The Date of Mark's Gospel Apart from the Temple and Rumors of War: The Taxation Episode (12:13–17) as Evidence," *CBQ* 79 (2017): 422–437; Christopher B. Zeichmann, "Capernaum: A 'Hub' for the Historical Jesus or the Markan Evangelist?" *JSHJ* 15 (2017): 147–165; Zeichmann, "Loanwords."

2. William E. Arnal, "The Gospel of Mark as Reflection on Exile and Identity," in *Introducing Religion: Essays in Honor of Jonathan Z. Smith*, ed. Willi Braun and Russell T. McCutcheon (London: Equinox, 2008), 57–67.

3. Important examples of support for a military reading of the pericope: Horsley, *Jesus and Empire*, 100–103; Brian J. Incigneri, *The Gospel to the Romans: The Setting and Rhetoric of Mark's Gospel*, Biblical Interpretation 65 (Leiden: Brill, 2003), 190–194; Matthias Klinghardt, "Legionsschweine in Gerasa: Lokalkolorit und historischer Hintergrund von Mk 5,1–20," *ZNW* 98 (2007): 28–48; Markus Lau, "Die *Legio X Fretensis* und der Besessene von Gerasa: Anmerkungen zur Zahlenangabe „ungefähr Zweitausend" (Mk 5,13)," *Biblica* 88 (2007): 351–364; Hans Leander, *Discourses of Empire: The Gospel of Mark from a Postcolonial Perspective*, SemeiaSt 71 (Atlanta: SBL, 2013), 201–219; Warren Carter, "Cross-Gendered Romans and Mark's Jesus: Legion Enters the Pigs (Mark 5:1–20)," *JBL* 133 (2015): 139–155.

4. Zeichmann, "Loanwords."

5. A third argument with less merit is the ostensive use of military terminology through-out the pericope. For instance, Richard Horsley suggests that "in a series of military images, Legion is 'dismissed' [5:13; *epestrepsen*] to enter the 'troop' [5:11, 13; *agelē*] of swine, who then 'charge' [5:13; *hōrmēsen*] headlong down the slope as if into battle. . . ." Greek and verse citations added; Horsley, *Jesus and Empire*, 100; cf. Carter, "Cross-Gendered Romans."

6. John R. Donahue and Daniel J. Harrington, *The Gospel of Mark*, Sacra Pagina 2 (Collegeville: Liturgical, 2002), 166.

7. Seyoon Kim, *Christ and Caesar: The Gospel and the Roman Empire in the Writings of Paul and Luke* (Grand Rapids: Eerdmans, 2008), 120.

8. Richard A. Horsley, "'By the Finger of God': Jesus and Imperial Violence," in *Violence in the New Testament*, ed. Shelly Matthews and E. Leigh Gibson (New York: T&T Clark, 2005), 68. Cf. Richard A. Horsley, *Hearing the Whole Story: The Politics of Plot in Mark's Gospel* (Louisville: Westminster John Knox, 2001), 144–146. Horsley relies heavily on the theoretical work of Frantz Fanon for his argument.

9. Horsley, *Hearing the Whole Story*, 145. See the criticism in Troels Engberg-Pedersen, "Review of Horsley, *Hearing the Whole Story*," *Journal of Theological Studies* 54 (2003): 244–245.

10. Cf. Leander, *Discourses of Empire*, 208–209.

11. See the discussion in E. Mary Smallwood, *The Jews under Roman Rule from Pompey to Diocletian: A Study in Political Relations*, SJLA 20, 2nd ed. (Leiden: Brill, 1981), 28–29; Schürer, *History*, 2.126.

12. William E. Arnal and Russell T. McCutcheon, *The Sacred Is the Profane: The Political Nature of "Religion"* (Oxford: Oxford University Press, 2013), 165.

13. Aleksander R. Michalak, *Angels as Warriors in Late Second Temple Jewish Literature*, WUNT II 330 (Tübingen: Mohr Siebeck, 2012), passim, esp. 86–98.

14. Arnal and McCutcheon, *Sacred Is the Profane*, 164. Their argument builds upon the insights of Luther H. Martin, *Hellenistic Religions: An Introduction* (Oxford: Oxford University Press, 1987).

15. Pheme Perkins, "The Gospel of Mark: Introduction, Commentary and Reflections," in *New Interpreters Bible* (Nashville: Abingdon, 1995), 8:584.

16. Misgav Har-Peled, "The Dialogical Beast: The Identification of Rome with the Pig in Early Rabbinic Literature" (Ph.D. dissertation, Johns Hopkins University, 2013), 130–131. Lightly edited for clarity.

17. Schürer, *History*, 1.371 n. 84 cites Seneca *Ira* 1.18.4, *Ben.* 3.25; Firmicus Maternus *Math.* 8.26.6; Dig. 48.20.6; Tg. Neof. Gen. 37:36, 40:3–4, 41:10–13.

18. John S. Kloppenborg, "*Evocatio Deorum* and the Date of Mark," *JBL* 124 (2005): 419–450.

19. W. A. Such, *The Abomination of Desolation in the Gospel of Mark: Its Historical Reference in Mark 13:14 and Its Impact in the Gospel* (Lanham: University Press of America, 1999), 170.

20. Incigneri, *Gospel to the Romans*, 133. Emphasis in original.

21. In a compelling article, Fernando Bermejo-Rubio argues that Mark and the Synoptic tradition more broadly attempts to alleviate Pilate's responsibility for Jesus' execution by minimizing the role of his personal staff and instead emphasizing the threat of the Jewish people. Pilate has no military informers, legal counsel, etc. in Mark or the other Gospels ("Was Pontius Pilate a Single-Handed Prefect? Roman Intelligence Sources as a Missing Link in the Gospel's Story," *Klio* 101 (2019): forthcoming).

22. For an excellent study demonstrating our nearly complete ignorance about the Roman practice of scourging, see Andrea Nicolotti, "The Scourge of Jesus and the Roman Scourge," *JSHJ* 15 (2017): 1–59.

23. John Granger Cook, *Crucifixion in the Mediterranean World*, WUNT I 327 (Tübingen: Mohr Siebeck, 2014), 428 n. 59.

24. Important examples understanding the passage as a taunt: Earl S. Johnson, "Is Mark 15.39 the Key to Mark's Christology?" *JSNT* 31 (1987): 3–22; Earl S. Johnson, "Mark 15:39 and the So-Called Confession of the Roman Centurion," *Biblica* 81 (2000): 406–413; Horsley, *Hearing the Whole Story*, 252; Donald Juel, "The Strange Silence of the Bible," *Interpretation* 51 (1997): 5–19; Nathan Eubank, "Dying with Power: Mark 15:39 from Ancient to Modern Interpretation," *Biblica* 85 (2014): 247–268. See other noteworthy publications cited in Kelly R. Iverson, "A Centurion's "Confession": A Performance-Critical Analysis of Mark 15:39," *JBL* 130 (2011): 349 n. 73.

25. Johnson, "Is Mark 15:39," 12–13. Johnson himself does not explicitly commit to a reading of mockery, but it seems to be implied here and in his follow-up article: Johnson, "So-Called Confession."

26. Stoll, "Religions of the Armies," 464–471; Hoey, "Official Policy." Cf. the *locus classicus* in Tacitus *Germ.* 43. There is also the Egyptian archive of Claudius Terentianus (enlisted in *classis Alexandriae* 110–136 CE; *P.Mich.* 8.476–481). In the letters between Claudius and his friends, cultic vocabulary is regularly invoked, but Roman deities are never named; Claudius and his friends instead opt for the god Serapis and associated deities. Likewise, the soldiers of *legio I Minervia* located in Germania Inferior had no interest in the Roman pantheon, instead worshipping local goddesses such as the Matronae Aufaniae: Haensch, "Inschriften."

27. Allen Brent, *The Imperial Cult and the Development of Church Order: Concepts and Images of Authority in Paganism and Early Christianity before the Age of Cyprian*, Supplements to Vigiliae Christianae 45 (Leiden: Brill, 1999), 255–265; Olivier J. Hekster, *Emperors and Ancestors: Roman Rulers and the Constraints of Tradition*, Oxford Studies in Ancient Culture and Representation (Oxford: Oxford University Press, 2015), 266–275. Emperors claimed to be the child of a single pantheon, but permitted other cults to identify them with other gods—especially Egyptian deities (continuous with traditions of both *Isis polyōnymos* and *interpretatio romana*). E.g., Domitian as son of Harachte (obelisk at Piazza Navona), Domitian as son of Isis, Osirus, and Re (Günther Hölbl, *Altägypten im Römischen Reich: der römische Pharao und seine Tempel. III. Heiligtümer und religiöses Leben in den ägyptischen Wüsten und Oasen* [Mainz: Zabern, 2005], 59).

Vespasian was reportedly the first emperor to encourage provincials to identify him as the son of a deity other than Venus. Though sources do not explicitly say Vespasian was claimed to be Horus incarnate (and thus the son of Isis and Serapis), he clearly cultivated a myth around the matter. Vespasian's self-mythologizing with Isis and Serapis should be understood as similar to his self-mythologizing with other Eastern cults as part of a propaganda campaign, including Judaism (Josephus), Ba'al Carmelus (oracle at Mount Carmel), the cult of Ammon (*P.Fouad* 8), and Cyprus' cult of Venus (oracle at Paphos). Incorporation of foreign cults into the imperial myth did not entail their incorporation into the Roman pantheon, nor did it entail compulsory recognition of emperor divinity within these cults.

There was diverse religious activity in the postwar garrison: amulets, inscriptions, and other cultic items have been found in Judaean military assemblages for numerous deities 66–135 CE, including Isis, Kore, Neotera, Serapis, Harpocrates, Eros, and Jupiter. Guy D. Stiebel, "Military Dress as an Ideological Marker in Roman Palestine," in *Dress and Ideology: Fashion-*

ing Identity from Antiquity to the Present, ed. Shoshana-Rose Marzel and Guy D. Stiebel (London: Bloomsbury, 2015), 160; Emmanuel Friedheim, "The Religious and Cultural World of Aelia Capitolina: A New Perspective," *Oriental Archive* 75 (2007): 125–152.

28. Stoll, "Religions of the Armies," 466.

29. Kloppenborg, *"Evocatio Deorum."*

30. Whitney T. Shiner, "The Ambiguous Pronouncement of the Centurion and the Shrouding of Meaning in Mark," *JSNT* 22 (2000): 9–11. Cf. Raymond E. Brown, *The Death of the Messiah. From Gethsemane to the Grave: A Commentary on the Passion Narratives in the Four Gospels*, Anchor Bible Reference Library, 2 vols. (New York: Doubleday, 1994), 1144–1145, helpfully distinguishing between historical readings (e.g., the temple veil's non-visibility from Golgotha) and literary readings (e.g., the rending of the temple veil as part of Mark's narrative).

31. Philo *Prov.* 2.46 is quoted in Eusebius *Praep. ev.* 8.14 and translated from Dale C. Allison, Jr., *Studies in Matthew: Interpretation Past and Present* (Grand Rapids: Baker, 2005), 95–96 n. 63. Allison and other commentators cite unnatural darkness at the deaths of Julius Caesar (Vergil *Georg.* 1.463–468; Pliny the Elder *Nat.* 2.30; Josephus *A.J.* 14.309), Augustus (Cassius Dio 56.29.3), Romulus (Cicero *Rep.* 6.21–22; Dionysius of Halicarnassus *Ant. rom.* 2.56; Ovid *Fast.* 485–498; Plutarch *Rom.* 27; Florus *Epit.* 1.1), Carneades (Diogenes Laertius 4.64), Pelopidas (Plutarch *Pel.* 295a), Proculus (Marinus *Proc.* 37), Enoch (2 En. 67.1–2), Adam (T. Adam 3.6), Matthias the high priest (Josephus *A.J.* 17.167), and Theodosius I (Ambrose *Ob. Theo.* 1). This is not to diminish the significance of Mark's allusion to Amos 8:9, merely to say Amos' symbolism was primed to be interpreted in this particular way in Roman antiquity.

32. Philip H. Bligh, "A Note on Huios Theou in Mark 15:39," *Expository Times* 80 (1968): 53. Emphasis in original. more recently, see David Álvarez Cineira, "The Centurion's Statement (Mark 15:39): A *restitutio memoriae*," in *Jesus—Gestalt und Gestaltungen: Rezeptionen des Galiläers in Wissenschaft, Kirche und Gesellschaft. Festschrift für Gerd Theißen zum 70. Geburtstag*, ed. Petra von Gemünden, David G. Horrell, and Max Küchler, Novum Testamentum et Orbis Antiquus 100 (Göttingen: Vandenhoeck & Ruprecht, 2013), 146–161.

33. Contrary to Bligh's claim about Jewish exception from the imperial cult, Monika Bernett, "Roman Imperial Cult in the Galilee," in *Religion, Ethnicity, and Identity in Ancient Galilee: A Region in Transition*, ed. Jürgen Zangenberg, Harold W. Attridge and Dale B. Martin, WUNT I 210 (Tübingen: Mohr Siebeck, 2007), 341 notes that "no law requiring veneration of the Roman emperor in cultic forms existed, and therefore nobody could be exempted from it." Indeed, there is no evidence of the imperial cult at all in the neighboring kingdom of Nabataea!

34. Shiner, "Ambiguous Pronouncement."

35. See the excellent study in S. R. F. Price, "Gods and Emperors: The Greek Language of the Roman Imperial Cult," *Journal of Hellenic Studies* 104 (1984): 79–95. Instances of *huios theou* (which Price notes was quite distinct from *divi filius*, relating to different conceptual systems for the Greek and Latin languages and their cultic formations) for emperors include *IG* 7.2713; *IGR* 3.286, 4.201, 4.309–311, 4.314, 4.594, 4.1302. See also John Granger Cook, *Roman Attitudes Toward the Christians: From Claudius to Hadrian*, WUNT I 261 (Tübingen: Mohr Siebeck, 2010), 29–37.

36. See the extensive discussion in Stephen Mitchell, "The Cult of Theos Hypsistos between Pagans, Jews and Christians," in *Pagan Monotheism in Late Antiquity*, ed. Polymnia Athanassiadi and Michael Frede (Oxford: Oxford University Press, 1999), 81–146. The

epithet was also commonly used of Ba'al in Syria, the Mother Goddess in Lydia, and Isis in Egypt, as well as occasional applications to Sabazios, Men, Attis, Poseidon, Eshmun, Eshmun-Melkart, and perhaps Helios: Paul Trebilco, *Jewish Communities in Asia Minor*, SNTSMS 69 (Cambridge: Cambridge University Press, 1991), 128–129, 239 n. 8.

37. Trebilco, *Jewish Communities*, 129. As examples, Trebilco cites T.Ash. 5.4; Jos. Asen. 8.2, 17.5; Josephus *A.J.* 16.163. One could also include Gen 14:18–20 LXX; Num 24:16 LXX; 2 Macc 3:31; 3 Macc 7:9; 1 Esd 2:2, 6:30, 8:19, 8:21; Dan 3:93, 4:2, 4:17, 4:34, 5:21 LXX-TH; Dan 3:93, 4:34, 4:37 LXX-OG; Esth 8E:17 LXX-OG; Acts 16:17; Philo *Legat.* 157, 317, *Leg.* 3.82.

38. Although historically there were many Jews in the Judaean *auxilia* before the Jewish War, Mark understands the soldiers at the cross to be Gentiles, since one passion prediction refers to the son of man's executors as "Gentiles" (10:32–34).

39. See also Raymond Brown's connection of Jesus' cry of God's abandonment with the centurion's acknowledgement of Jesus as God's son: Brown, *Death of the Messiah*, 1144.

40. R. T. France, *The Gospel of Matthew*, New International Greek Testament Commentary (Grand Rapids: Eerdmans, 2007), 84. Given the size of Bethlehem at the turn of the era, France estimates that there would have been roughly twenty boys executed.

41. *Ag.Ap.* 2.202. Jewish criticism of infanticide is very well documented, see the discussions in Erkki Koskenniemi, *The Exposure of Infants among Jews and Christians in Antiquity*, Social World of Biblical Antiquity II 4 (Sheffield: Sheffield Phoenix, 2009); Daniel R. Schwartz, "Did the Jews Practice Infant Exposure and Infanticide in Antiquity?" *Studia Philonica* 16 (2004): 61–95.

42. Richard A. Horsley, *The Liberation of Christmas: The Infancy Narratives in Social Context* (London: Continuum, 1989), 49.

43. Josephus *A.J.* 15.409; *J.W.* 1.282. See the helpful and nearly complete collection of relevant inscriptions in Richardson, *Herod*, 203–213. Richardson is critical of those who read the Masada amphorae as anything other than *regi herodi iudaico* (Mas 795–796, 800–801, 804–818, 821–826, 850, 946–950).

44. Laurena Ann Brink, "Going the Extra Mile: Reading Matt 5:41 Literally and Metaphorically," in *The History of Religions School Today: Essays on the New Testament and Related Ancient Mediterranean Texts*, ed. Thomas R. Blanton, IV, Robert Matthew Calhoun, and Clare K. Rothschild, WUNT I 340 (Tübingen: Mohr Siebeck, 2014), 111–128. Brink ultimately dismisses this sense of the word in favor of a more colloquial usage found in Jewish writings, though the writings she cites were composed centuries after Matthew and in different languages.

45. For scholarly reconstructions of the Q version, see Steven R. Johnson, ed., *Q 7:1–10: The Centurion's Faith in Jesus' Word*, Documenta Q (Leuven: Peeters, 2002).

46. On ethical-normative problems with this interpretation, see Christopher B. Zeichmann, "Rethinking the Gay Centurion: Sexual Exceptionalism, National Exceptionalism in Readings of Matt 8:5–13/Luke 7:1–10," *Bible and Critical Theory* 11/1 (2015): 35–54; Christopher B. Zeichmann, "Gender Minorities In and Under Roman Power: Respectability Politics in Luke–Acts," in *Luke-Acts*, ed. James Grimshaw, Texts@Contexts (London: Bloomsbury, 2018), 61–73. To summarize the primary objections: 1) one should not romanticize a relationship between a military officer and a slave, given the latter's inability to consent, 2) the interpretation of the pericope often relies upon "respectability politics" that locates the dignity of queers in their compliance with state violence, and 3) this interpretation often works on racist assumptions about the perversity of the ethnic "other" that misrepresents Jew-

ish sexual norms of the early Roman period. The point, to be clear, is not that queer biblical interpretation should not be done, but that *it can be done better* via serious engagement with intersectionality.

47. Theodore W. Jennings, Jr. and Tat-Siong Benny Liew, "Mistaken Identities but Model Faith: Rereading the Centurion, the Chap, and the Christ in Matthew 8:5–13," *JBL* 123 (2004): 467–494.

48. Saddington, "Centurion in Matthew."

49. For more on the social structures linking military sexual norms to Rome's imperial ambitions, see David J. Mattingly, *Imperialism, Power, and Identity: Experiencing the Roman Empire* (Princeton: Princeton University Press, 2011), 94–121; Whittaker, *Rome and Its Frontiers*, 115–143; Zeichmann, "Rethinking the Gay Centurion," 48–50.

50. Zeichmann, "Rethinking the Gay Centurion," 50, citing §§119–120 and Josephus *A.J.* 19.357.

51. Dale B. Martin, "Jesus in Jerusalem: Armed and Not Dangerous," *JSNT* 37 (2014): 4–7.

52. See Paula Fredriksen, "Arms and the Man: A Response to Dale Martin's 'Jesus in Jerusalem: Armed and Not Dangerous,'" *JSNT* 37 (2015): 312–325; Francis Gerald Downing, "Dale Martin's Swords for Jesus: Shaky Evidence?" *JSNT* 37 (2015): 326–333.

53. John S. Kloppenborg, "The Representation of Violence in Synoptic Parables," in *Mark and Matthew I: Comparative Settings: Understanding the Earliest Gospels in Their First Century Settings*, ed. Eve-Marie Becker and Anders Runesson, WUNT I 271 (Tübingen: Mohr Siebeck, 2011), 351.

54. See the discussion in Graham Sumner, *Roman Military Clothing (1): 100 BC AD 200*, Men-at-Arms 374 (Oxford: Osprey, 2002), 16–17.

55. Sumner, *Roman Military Clothing*, 34–35.

56. Dorothy Jean Weaver, "'Thus You Will Know Them by Their Fruits': The Roman Characters of the Gospel of Matthew," in *The Gospel of Matthew in Its Roman Imperial Context*, ed. John Riches and David C. Sim, JSNTSup 276 (London: T&T Clark International, 2005), 114.

57. Still others read Luke-Acts as a text that resists the Roman Empire: e.g., Amanda C. Miller, *Rumors of Resistance: Status Reversals and Hidden Transcripts in the Gospel of Luke*, Emerging Scholars (Minneapolis: Fortress, 2014).

58. Ephesus: Christopher B. Zeichmann, "οἱ στρατηγοὶ τοῦ ἱεροῦ and the Location of Luke–Acts' Composition," *Early Christianity* 3 (2012): 172–187; Richard I. Pervo, *Acts: A Commentary*, Hermeneia (Minneapolis: Fortress, 2009), 5–7; Jan Lambrecht, "Paul's Farewell-Address at Miletus: Acts 20, 17–38," in *Les Actes des Apôtres: Traditions, Rédaction, Théologie*, ed. Jacob Kremer, BETL 48 (Leuven: Leuven University Press, 1979), 307–337. 115 CE: Richard I. Pervo, *Dating Acts: Between the Evangelists and the Apologists* (Sonoma: Polebridge, 2006).

59. Seth Schwartz, *Josephus and Judaean Politics*, Columbia Studies in the Classical Tradition 18 (Leiden: Brill, 1990), 97 n. 137.

60. Roth, "Jews and the Roman Army," 413.

61. See *P.Fouad* 1.21 (63 CE), *P.Yale Inv.* 1528 and their discussion in Denis B. Saddington, *The Development of the Roman Auxiliary Forces from Caesar to Vespasian (49 B.C.–A.D. 79)* (Harare: University of Zimbabwe Press, 1982), 65, 189.

62. Zissu and Ecker, "Roman Military Fort," 301.

63. Though biblical commentators tend to cite opinions favorable to their interpretation, the existence of a first-century synagogue in Capernaum is uncertain among archaeologists. E.g., Dennis E. Groh, "The Stratigraphic Chronology of the Galilean Synagogue from the

Early Roman Period Through the Early Byzantine Period (ca. 420 C.E.)," in *Ancient Synagogues: Historical Analysis and Archaeological Discovery*, ed. Dan Urman and Paul V. M. Flesher, Studia Post-Biblica 47, 2 vols. (Leiden: Brill, 1998), 1.57: "insufficient evidence to stand as [a] first-century synagogue."

64. E.g., Domination: Joseph A. Fitzmyer, *The Gospel of Luke: A New Translation and Commentary*, Anchor Bible Commentary 28A (Garden City: Doubleday, 1985), 1346–1347. Mission: Vittorio Fusco, "Problems of Structure in Luke's Eschatology Discourse (Luke 21:7–36)," in *Luke and Acts*, ed. Gerald O'Collins and Gilberto Marconi (New York: Paulist, 1991), 87. Both: Joel B. Green, *The Gospel of Luke*, New International Commentary on the New Testament (Grand Rapids: Eerdmans, 1997), 739. Other: John Nolland, "'The Times of the Nations' and a Prophetic Pattern in Luke 21," in *Biblical Interpretation in Early Christian Gospels*, ed. Thomas R. Hatina, Library of New Testament Studies 376, 5 vols. (London: T&T Clark, 2010), 3.146–147.

65. Arnal, *Jesus and the Village Scribes*, 172–183.

66. Zeichmann, "οἱ στρατηγοὶ τοῦ ἱεροῦ."

67. See the classic discussion in Adrian N. Sherwin-White, *Roman Society and Roman Law in the New Testament* (Oxford: Oxford University Press, 1963), 28–31. That is, the practice of *forum delicti* (trial at the place of crime) was used in the time of Jesus and *forum domicilii* (trial at the place of residence) was practiced later.

68. See, e.g., John S. Kloppenborg, "*Exitus clari viri*: The Death of Jesus in Luke," *Toronto Journal of Theology* 8 (1992): 106–120.

69. Brink, *Soldiers in Luke-Acts*, 106–109.

70. Michael P. Speidel, "The Roman Army in Judaea under the Procurators: The Italian and the Augustan Cohort in the Acts of the Apostles," *Ancient Society* 13–14 (1982–83): 233–240; cf. Schürer, *History*, 1.365; Zeichmann "Military Forces."

71. For collections of early Christian texts discussing the military, see George Kalantzis, *Caesar and the Lamb: Early Christian Attitudes on War and Military Service* (Eugene: Cascade, 2012); Ronald J. Sider, *The Early Church on Killing: A Comprehensive Sourcebook on War, Abortion, and Capital Punishment* (Grand Rapids: Baker Academic, 2012), 137–162. See also the excellent study in John F. Shean, *Soldiering for God: Christianity and the Roman Army*, History of Warfare 61 (Leiden: Brill, 2010).

72. Edward M. Peters, "Prison Before the Prison: The Ancient and Medieval Worlds," in *The Oxford History of the Prison: The Practice of Punishment in Western Society*, ed. Norval Morris and David J. Rothman (Oxford: Oxford University Press, 1995), 16–17.

73. Richard J. Cassidy, *Paul in Chains: Roman Imprisonment and the Letters of St. Paul* (New York: Crossroad, 2001); Brian Rapske, *The Book of Acts and Paul in Roman Custody*, The Book of Acts in Its First Century Setting 3 (Grand Rapids: Eerdmans, 1994); Craig S. Wansink, *Chained in Christ: The Experience and Rhetoric of Paul's Imprisonments*, JSNTSup 130 (Sheffield: Sheffield Academic, 1996). But see Jens-Uwe Krause, *Gefängnisse im römischen Reich*, Heidelberger althistorische Beiträge und epigraphische Studien 23 (Stuttgart: Steiner, 1996).

74. Saundra Schwartz, "The Trial Scene in the Greek Novels and in Acts," in *Contextualizing Acts: Lukan Narrative and Greco-Roman Discourse*, ed. Todd Penner and Caroline Vander Stichele, SBLSS 20 (Atlanta: SBL, 2003), 120–121.

75. On the role of soldiers as watchmen, see Krause, *Gefängnisse*, 252–254.

76. See the classic discussion in Wolfgang Stegemann, "War der Apostel Paulus ein römischer Bürger?" *Zeitschrift für die neutestamentliche Wissenschaft* 78 (1987): 200–229. More recently, see Karl-Leo Noethlichs, "Der Jude Paulus: Ein Tarser und Römer?" in *Rom und*

das himmlische Jerusalem: Die frühen Christen zwischen Anpassung und Ablehnung, ed. Raban von Haehling (Darmstadt: Wissenschaftliche Buchgesellschaft, 2000), 67–84; Ekkehard W. Stegemann and Wolfgang Stegemann, *The Jesus Movement: A Social History of Its First Century*, trans. O. C. Dean, Jr. (Minneapolis: Fortress, 1999), 297–302; Harry W. Tajra, *The Trial of St. Paul: A Juridical Exegesis of the Second Half of the Acts of the Apostles* (Eugene: Wipf and Stock, 2010).

77. H. Devijver, "Equestrian Officers from the East," in *Defence of the Roman and Byzantine East*, ed. David L. Kennedy and Philip Freeman, BARIS 297, 2 vols. (Oxford: BAR, 1986), 1.109–225; H. Devijver, "Equestrian Officers in the East," in *The Eastern Frontier of the Roman Empire*, ed. David H. French and Chris S. Lightfoot, BARIS 553, 2 vols. (Oxford: BAR, 1989), 1.77–111; H. Devijver, "The Geographical Origins of Equestrian Officers," in *The Equestrian Officers of the Roman Imperial Army*, ed. H. Devijver, Marvors Roman Army Researches 9, 2 vols. (Stuttgart: Steiner, 1992), 2.109–128.

78. Devijver, "Equestrian Officers in the East." He draws upon Eric Birley, *Roman Britain and the Roman Army: Collected Papers*, 2nd ed. (London: Kendal, 1961), 147–148.

79. Zeichmann, "Gender Minorities." This is not to downplay the deeply negative depiction of the Judaean governor Felix in Acts; see Joshua Yoder, *Representatives of Roman Rule: Roman Provincial Governors in Luke-Acts*, Beihefte zur Zeitschrift für die neutestamentliche Wissenschaft 209 (Berlin: De Gruyter, 2014), 277–302.

80. Hans-Josef Klauck, *Ancient Letters and the New Testament: A Guide to Context and Exegesis*, trans. Daniel P. Bailey (Waco: Baylor University Press, 2006), 433.

81. Most scholars draw upon the influential discussion in Speidel, "Roman Army in Judaea," 237–240. But see Zeichmann, "Military Forces"; Zeichmann, "Herodian Kings," the latter requiring updating due to subsequently discovered inscriptions. On Agrippa II in the aborted Parthian campaign: Tacitus *Ann.* 13.7. On the Jewish War: Josephus *J.W.* 2.500, 3.68.

82. Applebaum, *Judaea*, 57.

83. Christine R. Shea, "Names in Acts: A Cameo Essay," in *Acts and Christian Beginnings: The Acts Seminar Report*, ed. Dennis E. Smith and Joseph B. Tyson (Salem: Polebridge, 2013), 22–24.

84. Dennis E. Smith and Joseph B. Tyson, ed., *Acts and Christian Beginnings: The Acts Seminar Report* (Salem: Polebridge, 2013), 120. While the precise date of Rome's foundation was disputed during the Republic, a consensus emerged during the empire: 21 April 753 BCE. Recall also that Julius Caesar and *gens Iulia* more broadly claimed descent from Aeneas' son Iulus, the mythological firstborn of the Roman line.

85. See Steven J. Friesen, "Poverty in Pauline Studies: Beyond the New Consensus," *JSNT* 26 (2004): 323–361; Friesen, "Paul and Economics."

86. Richard I. Pervo, *Profit with Delight: The Literary Genre of the Acts of the Apostles* (Philadelphia: Fortress, 1987), 45–46.

87. See the extended studies in Brink, *Soldiers in Luke-Acts*; Kyrychenko, *Roman Army*.

88. Kyrychenko, *Roman Army*, 189.

89. There is also a compelling interpretation of the Parable of the Wicked Manager (Luke 19:11–27) that is a thinly veiled polemic against the Herodian ethnarch Archelaus; see a similar interpretation in Brian Schultz, "Jesus as Archelaus in the Parable of the Pounds (Lk. 19:11–27)," *NovT* 49 (2007): 105–127. I wish to thank Jordan Balint for helping me with the role of Herodian kings in Luke-Acts.

90. Simon James, "The Community of the Soldiers: A Major Identity and Centre of Power in the Roman Empire," in *Proceedings of the Eighth Theoretical Roman Archaeology*

Conference, ed. Patricia Baker, Colin Forcey, Sophia Jundi, and Robert Witcher (Oxford: Oxbow, 1999), 14; cf. Simon James, "Soldiers and Civilians: Identity and Interaction in Roman Britain," in *Britons and Romans: Advancing an Archaeological Agenda*, ed. Simon James and Martin Millett, Council for British Archaeology Research Report 125 (York: Council for British Archaeology, 2001), 77–89.

91. See Robert T. Fortna, *The Fourth Gospel and Its Predecessor: From Narrative Source to Present Gospel* (Philadelphia: Augsburg Fortress, 1988); Burton L. Mack, *A Myth of Innocence: Mark and Christian Origins* (Philadelphia: Fortress, 1988), 220–230; Burton L. Mack, *Who Wrote the New Testament? The Making of the Christian Myth* (San Francisco: HarperSanFrancisco, 1995), 176–183.

92. I understand the first stage to comprise 4:45a, 46a, 47, 49–52. The remaining portion of John 4:43–54 corresponds to the second stage.

93. Zeichmann, "Capernaum," 162–163.

94. See, e.g., the inscriptions discussed in Anthony R. Birley, *The People of Roman Britain* (Berkeley: University of California Press, 1980), 148.

95. E.g., §13, *IG* 2.2.3441; *OGIS* 419, 420, 424. See the discussion in David Braund, *Rome and the Friendly King: The Character of Client Kingship* (London: Croom Helm, 1984), 105–111.

96. Albert Wifstrand, "Autokrator, Kaisar, Basileus," in *ΔPAΓMA: Martino P. Nilsson, A.D. IV id. iul. anno MCMXXXIX dedicatum*, ed. H. Ohlsson, Skrifter Utgivna av Svenska Institutet i Rom 1 (Lund: Ohlssons, 1939), 529–539; cf. E. A. Judge, *The First Christians in the Roman World: Augustan and New Testament Essays*, WUNT I 229 (Tübingen: Mohr Siebeck, 2008), 395–403.

97. Giovanni B. Bazzana, *Kingdom of Bureaucracy: The Political Theology of Village Scribes in the Sayings Gospel Q*, Bibliotheca ephemeridum theologicarum lovaniensium 274 (Leuven: Peeters, 2015), 213–262. But contrast, e.g., Josephus *J.W.* 5.563.

98. Michelle Christian, "Calculating Acts: Luke's Numeracy in Context," in *Luke on Jesus, Paul, and Christianity: What Did He Really Know?*, ed. John S. Kloppenborg and Joseph Verheyden, Biblical Tools and Studies 29 (Leuven: Peeters, 2017), 219–239.

99. Brown, *Death of the Messiah*, 1075–1077.

100. G. D. Kilpatrick, "The Transmission of the New Testament and Its Reliability," *Bible Translator* 9 (1958): 133–134.

101. Bruce M. Metzger, *A Textual Commentary on the Greek New Testament*, 2nd ed. (New York: United Bible Societies, 1994), 217–218.

102. Brian Johnson, "Crurifragium: An Intersection of History, Archaeology, and Theology in the Gospel of John," in *My Father's World: Celebrating the Life of Reuben G. Bullard*, ed. John D. Wineland, Mark Ziese, and James Riley Estep (Eugene: Wipf and Stock, 2011), 92–95.

4

The Military in the Pauline Corpus

The purpose of this chapter is conventional in method and limited in scope, offering a contrast between the voice of the "authentic" Pauline epistles and the "disputed" voices writing in his name on military matters. Whereas the "authentic" epistles use the military as one floating metaphor among others to speak of unrelated matters, in the "inauthentic" writings—most notably Ephesians and the Pastoral Epistles—the military functions as a Roman-friendly model for what early (Pauline) Christianity should be. This discrepancy may be most evident in Paul's apparent unfamiliarity with the Roman military (e.g., provisioning in 1 Cor 9:7) and Ephesians' highly spiritual understanding of early Christianity as a project in deep harmony with the Roman military imagination. Thus, while both the authentic and disputed letters discuss spiritual armor, the purpose is quite different between the two; in the former, it is an incidental image invoked in service of an unrelated point, while in the disputed epistles it bears a crucial relationship to the social project that the author advances.

The scope of this chapter is limited, as several Pauline texts that scholars might consider relevant are not discussed here, including Paul's discussion of principalities and powers (e.g., 1 Cor 2:8, Rom 8:37–39), passing references to battles within (e.g., Rom 7:23; cf. 1 Pet 2:11, Jas 4:1), nonmilitarized language of battles (e.g., 1 Thes 2:2), and his discussion of civil authorities (Rom 13:1–7). The reasons for these omissions are multitude: they do not directly refer to the military, scholarly discussion on either issue requires a degree of detail they cannot be afforded here, uncertainty surrounding their meaning and significance, etc. Many key terms that formerly bore overtly military connotations (e.g., *agōn*, *sōtēr*) were imbued with less martial significance when Paul and his successors wrote, operating with a much broader semantic domain; *agōnizomai* characterizes combat in the famous dispute between Ajax and Odysseus over Achilles' armor (Homer *Il.* 23.706–739), but merely means something like "struggle" or "contest" in the Pauline corpus (e.g., 1 Cor 9:25, Col 1:29). When focusing upon

Paul's more explicit military language, we will see that his interest in the topic does not extend far beyond the incidental metaphor, despite occasional suggestions that the military figures deeply into his thinking (e.g., Army of Light vs. Army of Darkness).[1] As with the preceding chapter, this is not a thorough exegetical commentary exploring all possible meanings, but serves to observe aspects in *corpus paulinum* that may be illuminated by discussion of the military.

Though nearly the entire second half of Acts of the Apostles is devoted to the vicissitudes of Paul's ministry, that depiction of the apostle will not inform the discussion in this chapter. First, there is reason to be suspicious of Acts' historical reliability. Acts has a distinct set of narrative interests that render its use in reconstructing the life of Paul too complicated for use here. For instance, the dialogue of Paul in Acts sounds more like the author of Acts than the Pauline epistles. Moreover, Acts contradicts the letters of Paul at times. Paul, for instance, claimed to be uneducated in his letters (2 Cor 11:6), but Acts' character boasts about learning at the feet of Gamaliel (22:3). One gets the impression from Paul's letters that he mostly ministered to manual laborers and the impoverished, whereas in Acts he spends most of the time talking to society's elite.[2] Acts and the epistles disagree on many other points, some of which will be discussed below (e.g., his prison experience). Second, despite the tensions between Acts' depiction of Paul and what one finds in his own letters, Paul's letters were likely used in the composition of Luke-Acts' two-volume Christian epic.[3] Scholars have thus argued that Acts was familiar with the letters of Paul and reinterprets key moments in Paul's life, a reading tendency that often obscures Paul's distinctive voice in the history of interpretation. That is, there is a tendency to harmonize Paul's letters with the narrative in Acts, even if it is unintentional. Third, Acts is simply not part of the Pauline corpus and does not claim to be written by Paul. For reasons of consistency, it is prudent to treat these as discrete literary units.

Methodologically "forgetting" Acts' Paul is not an easy task due to Acts' interpretive inertia for our default image of early Christianity. Many important aspects of Paul's life are only attested in Acts and historically specious: the name Saul, his origin in Tarsus, his Roman citizenship, his education, his literacy, the notion of a "conversion" distinct from a "calling," Paul having any control over his travels, and so on. This is not to mention how several events attested in the epistles are significantly recast in Acts: escape from a Nabataean ethnarch becomes part of Acts' schema of Jews persecuting Paul, his relationships with other early Christians move from an agree-to-disagree policy to an amicable doctrinal agreement, Paul's horrific incarceration becomes house arrest with a friendly centurion, etc. Pauline biography is outside the scope of this book, but it will be addressed as relevant here. My own tendency is to treat Acts' depiction of Paul as somewhat comparable to the genre of "historical fiction" today: a largely fictional narrative based around a historical figure and a handful of key episodes in history that serves to entertain as much as it does edify. This study assumes Acts is not a reliable source for reconstructing Paul's life.

Finally, due to the material repeated in the letters, this chapter will be more thematic in its treatment than was found in the discussion of the Gospels. This approach also accounts for the complicated way in which many of the canonical epistles are composite documents, collecting several letters into a single book. This chapter will proceed in a roughly chronological manner, both within the Pauline corpus and from the "authentic" to the "disputed" letters. These terms are left in scare quotes, as the notion of authentic authorship is a theoretical problem that requires further attention.[4]

"AUTHENTIC" PAUL

The Pauline corpus consists of fourteen letters, though scholars are unanimous that not all were composed by the apostle Paul—indeed, the book of Hebrews does not even claim to be written by Paul, even though early Christians designated it part of the corpus. I assume here the authenticity of seven "undisputed letters" (1 Thessalonians, 1–2 Corinthians, Philippians, Philemon, Galatians, and Romans—composed roughly in that order) and the pseudonymity of the seven "disputed letters" (Colossians, Ephesians, 2 Thessalonians, 1–2 Timothy, Titus, Hebrews). These will be treated in different sections within this chapter, as the present study concerns the distinctive voices of different authors on the matter of the military; even though the author of Ephesians was clearly influenced by Paul, it is prudent to treat the two as discrete authors. This difference between these voices is especially evident with the question of when these authors wrote: authentic Pauline letters were written in the 50s and 60s CE, when the apostle and others were forming new congregations in disparate parts of the Roman East; letters of doubtful authenticity were written considerably later, when it could be taken for granted that Christianity was predominantly Gentile, when Christians were less preoccupied with the coming Day of the Lord and more interested in how to go about living their lives in the Roman Empire, and so on. It will also become clear that this differing social setting played a significant role in their understandings of the military.

It is also at this point of the present study that we leave Palestine and most of the Levant. Though Paul himself spent time in Judaea (Gal 2:1–10), we have no letters Paul wrote to Palestinians, nor any letters that were likely written while he was in Palestine. Paul's surviving correspondence was, with the exception of Romans, entirely with residents of the Roman East, which seems to have been subject to some overarching military policies that were also found in Palestine.[5] Very few with whom Paul corresponded were Jews or Palestinians and had little reason to care much about Palestine. There were naturally many differences between individual provinces and cities, but much of the social-historical discussion from chapters 1 and 2 can be carried over here as well.

God's Imperial Parade—1 Thes 4:16–17; 1 Cor 15:20–28, 52; 2 Cor 2:14–16

In Paul's earliest letters, the apostle paints a brief, if memorable, portrait of the impending "Day of the Lord." When writing to the Thessalonians, Paul describes a scenario wherein God's descent from heaven onto earth is preceded by a signal on behalf of the Lord (*kyrios*), the voice of an archangel, and the sound of God's trumpet—an instrument that has military connotations elsewhere in his letters (1 Cor 14:8). A similar situation seems to be assumed in 1 Corinthians as well. As many commentators have noted, Paul describes this coming with the word *parousia*, a term associated with the presence of the imperial visitors, often bearing the honorific title *kyrios*, to a city.[6] One Egyptian administrator, for instance, writes a frantic letter concerning the additional work required by the king's imminent *parousia* at a nearby village (*P.Tebt.* 1.48). The term is well attested across the eastern Mediterranean several centuries before and after Paul wrote. Consequently, many have argued that Paul draws upon Greco-Roman imperial frameworks for his apocalyptic scenario.

But why does Paul adopt this imperial language? Is he simply using this language without reflection or does the metaphor of *parousia* serve a larger purpose in his writings? Scholars often suggest the latter, contending that Paul's Day of the Lord is anti-Roman polemic. Though these letters indicate that Paul believes God will rescue Christians from the impending wrath, such interpreters give it a political inflection: the Lord's *parousia*, not the emperor's, will save them from destruction. Indeed, the emperor himself will suffer the same grisly fate as the rest of creation. This reading finds support in Paul's criticism of those who assert there will be "peace and security" (5:1), rejecting the imperial slogans *pax et securitas* and *pax Romana*.[7] By deliberately deploying political terminology in his eschatological scenario, Paul locates "true" peace, security, and salvation in the apocalyptic appearance of Jesus and not the emperor's own claim of such.

Other commentators have objected to the idea that Paul was inspired by Greco-Roman imperialism in his eschatology. They commonly point to the Septuagint (2 Macc 8:12, 15:21) and other Jewish texts as more likely sources for language of *parousia* as a divine appearance: Josephus *A.J.* 3.80, 9.55; T.Jud. 22.3; T.Levi 8.15; cf. its use for Asclepius in *IG* 4.2.122.[8] There need not be a contradiction here, as it is likely the theophantic language of older Jewish literature was interpreted in light of the imperial usage. It is also worth noting that this terminology is first found in Jewish literature of the Greco-Roman period when it had imperial overtones.

Relating more specifically to the military, the *parousia* in Paul's letters may refer to an imperial victory procession: the Lord's signal, the cry from the archangel, and the sound of God's trumpet sound suspiciously like a military parade, specifically a triumph or ovation. The triumph was the single highest honor bestowed upon a general and granted at the behest of the Senate. There was no strict formula for a triumph's proceedings, but common elements included an early morning speech by the victorious general, salutes from the crowd, dispersal of gifts to the army, a procession through the city displaying war captives (soon to be sold into slavery or executed), and culmination in a sacrifice to Jupiter with a public feast. During Rome's imperial period, triumphs were limited to emperors. Less prestigious were ovations, which were granted for less decisive victories, for smaller conflicts (fewer than 5,000 casualties), or for victories over less "worthy" foes. Triumphs and ovations were uncommon and limited to the city of Rome, meaning very few in the Roman Empire ever saw one. That said, both varieties of parade loomed large in the popular imagination and were imitated on occasion, David Catchpole cites twelve examples from Judaea alone, including Alexander the Great (Josephus *A.J.* 11.325–339), Apollonius (2 Macc 4:21–22), Simon Maccabee (1 Macc 13:49–51), and Marcus Agrippa (Josephus *A.J.* 16.12–15).[9] Catchpole notes that these stories follow a fairly consistent formula: a) a recognized person achieves victory, almost always *military* victory; b) a formal ceremonial entrance; c) greetings or acclamation with an invocation of God; d) entry to the city climaxing with a visit to the temple; and e) cultic activity either positive or negative. Commentators have suggested that Jesus' entry into Jerusalem (Mark 11:1–11/parallels) echoes such parades and Paul explicitly invokes the imperial parade in 2 Cor 2:14, albeit with different connotations.[10] Paul also approximates this in his description of the Day of the Lord, mostly emphasizing the ceremony of the Lord's entrance and his greeting from both the dead in Christ and those still living.

Paul conflated two similar images in his eschatological scenario: that of the imperial arrival to a provincial city (i.e., *parousia*) and that of the victory parade. Paul is prone to mix metaphors throughout his letters; in this instance he draws attention to the majesty of the Lord's entrance at the expense of a single cohesive image. When Paul wrote these letters, he had not yet been to Rome and, from what we know of his travels, he was nowhere near any military skirmishes; like nearly everyone else in the Roman Empire, Paul had probably never seen a military victory parade or witnessed the emperor's *parousia*. Instead, he and others were largely familiar with these events from local lore and imperial propaganda (e.g., §382, figure 4.1). These events varied in their performance and their recollection in the historical record was

Figure 4.1.   Roman Imperial Triumph. Detail from the Arch of Titus (82 CE), depicting the triumph held in the eventual-emperor Titus' honor for his victory as a general in the Jewish War. It was erected in the city of Rome by the emperor Domitian, Titus's brother.
Photo credit: Xavier Espinasse, used by permission.

similarly mutable. Consequently, it is difficult to know what to make of the silences in Paul's imagery. How might we read the absence of prisoners-of-war, for example? Was Paul opposed to imprisoning one's enemies because of its brutality? Or are we to infer that people not rescued by the Lord suffer a grisly fate similar to Roman captives? The unappealing prospect of the latter may be implied in 1 Cor 15:24 and 15:27, which describe Christ as vanquishing *all* his foes—there is little room for ambiguity.[11] This would seem further suggested in 2 Cor 2:14–16, wherein Paul brings death as part of God's triumphal procession (*thriambeuonti*). Whether Paul is an active or passive participant in the triumph is debated, but Neil Elliott is no doubt correct in suggesting that this is an ironic reversal in Paul's mind: the Roman power that has publicly ridiculed him will have its own military rituals turned against it when the Day of the Lord arrives.[12]

Some version of Paul's apocalyptic scenario was already familiar to the Thessalonians, as his brief summary in 1:9–10 begins with the reminder that they already know some of these things. Paul's argument in 1 Cor 15 likewise presumes his readers' familiarity with the Lord's rescue of the faithful. For whatever reason, perhaps because he did not know himself, Paul did not feel the need to flesh out the specifics of the Day of the Lord in either of these letters.

Weapons of Spiritual Warfare—1 Thes 5:8; 2 Cor 6:7; Rom 6:13, 13:12

Paul frequently uses metaphors, occasionally touching upon weapons of spiritual warfare, metaphors he does not feel the need to elaborate. In his letter to the Thessalonians (5:4–11), Paul mixes a series of metaphors: watchfulness concerning the day of the Lord, sobriety of mind, and the metaphor of armor.[13] Paul describes a breastplate of faith and love and a helmet of the hope of salvation, drawing upon the imagery of Isa 59:17 (cf. Wis 5:17–20): "he put on righteousness as his breastplate and the helmet of salvation on his head." The choice is odd rhetorically, since the Thessalonians were Gentile and had only passing familiarity with the Hebrew Bible and Septuagint. Regardless, the image of armor was familiar in the Roman East, even if there is no evidence of a Roman garrison in the city of Thessalonica itself.

The image of divine armor appears again in Paul's epistle to the Romans after his discussion of obedience to authorities (13:12). This time, the apostle contrasts the time of darkness with the "armor of light," with no apparent biblical intertexts. The offhand reference suggests Paul was not particularly invested in the image; it would not have taken much work to integrate the metaphor into his discussion of worldly authorities only a few verses earlier, had it been an important symbol. This might be contrasted with pseudo-Pauline literature, which consistently connects its military imagery with a normative political stance toward Roman power.

In 2 Cor 6:7, Paul refers to armaments of righteousness held in both hands. There is some debate as to whether the use of both right and left hand indicates a defensive (i.e., both weapon and shield) or aggressive posture (i.e., two weapons). The latter is unlikely, as dual-wielding was a strange way of positioning one's self for combat. Dual-wielding weapons was not appropriate for military combat, though gladiators of the *dimachaerus* type used two swords in the arena, requiring a fighting style that one modern commentator ridicules as impractical and "inexplicable."[14] There is no reason a soldier would use two swords in combat, nor any reason a civilian like Paul would suppose it. The image of dual-wielding is largely attributable to more recent action movies set in antiquity, an image that sets aside any pretense of realism in the process. However, we need not suppose that use of both a weapon and shield indicates an inherently *defensive* stance; rather, it suggests a state of preparation for whatever threat arrives. This is similar to the weapons of righteousness and unrighteousness that Paul mentions in Rom 6:13. In these latter instances, the military images are mere throwaways—stock images that Paul assumed were familiar to his audience and which warranted no explanation.[15]

Soldiers' Pay—1 Cor 9:7

In one of his letters to the Corinthians, Paul defends his right to be paid as the Corinthians' apostle, comparing his situation with other workers compensated for their labor, including soldiers. The example of the soldier specifically concerns food: Paul asks, who serves as soldier and pays for their own rations (*opsōnion*)?[16] The force of Paul's argument implies that the answer is "no one." We might recall from chapter 1 that this does not reflect actual military practices, as food expenses were deducted from soldiers' pay. Provisions were a common point of contention in the military; food is a uniting concern among the numerous military letters found at Wadi Fawakhir in Egypt, most of which indicate soldiers paid for their own rations and foodstuffs beyond what was served in the garrison. One letter between soldiers verges on comedy in its anxiety about food:

> Rustius Barbarus to Pompeius, greetings. First of all, I pray that you are in good health. Why do you write me such a nasty letter? Why do you think I am so thoughtless? If you did not send me the green vegetables so quickly, must I immediately forget your friendship? I'm not like that, or thoughtless either. I think of you, not as a pal, but as a twin brother, the same flesh and blood. It's a term I give you quite often in my letters, but you think of me in a different light. I have received bunches of cabbage and one cheese. I have sent you by Arrianus, the trooper, a box, inside which is one cake and a denarius wrapped in a small cloth. Please buy me a *matium* of salt and send it to me without delay, because I want to bake some bread.[17]

Numerous other examples could be cited from the papyrological record: mothers send their enlisted sons food and are frustrated that they do not write back sooner, soldiers complain about the quality of dining available and ask friends to pick up better food at the market, receipts indicate the specific costs and eating habits of individual soldiers, etc.

Paul is misinformed in 1 Cor 9:7: at least with respect to food-rations, soldiers indeed served at their own expense. The apostle may have merely conflated the (correct) fact that soldiers do not fight at their own expense with the reasonable but incorrect inference that this extended to the costs of their own upkeep. Military receipts indicate that it was standard for rations to be deducted from soldiers' pay *before* they received their wages (e.g., §22), not to mention any additional food paid from a soldier's private coffers. Paul's comments indicate the gap between the lived life of soldiers and the perception of such lives among civilians. As with the imperial parade, this would be substantial evidence that Paul writes at considerable social distance from the military, operating on knowledge that seems to be inferential, incidental, legendary, and part of a blurry imaginary; this is not the writings of a person with substantial firsthand experience with the military. Though Paul considered it obvious that soldiers did not pay for their own food, he was incorrect in his assumption.

Paul's Siege Warfare—2 Cor 10:3–6; Rom 8:37

In one of the epistles comprising 2 Corinthians (partially preserved in 2 Cor 10:1–13:10), Paul uses an extended military metaphor to describe spiritual combat against demonic forces. Paul deploys a dichotomy here that recurs throughout his epistles to the Corinthians, that of flesh vs. Godly spirit, a distinction that seems to be a point of tension between Paul and the Corinthians. This metaphor is given special prominence: it prefaces Paul's extended defense of his ministry, rhetorically positioning the following material as part of an all-out war on his critics, implying they are opposed to Paul's alliance of godly forces. It was common for writers, orators, and philosophers of the Greco-Roman period to depict themselves in a figurative military battle against rivals (e.g., Seneca *Lucil.* 109.8–9, Seneca the Elder *Controv.* 9 pref. 4, Tacitus *Dial.* 34, 37, Diogenes of Sinope *Cyn. Ep.* 10, Diogenes Laertius 6.12–13, Epiphanius *Pan.* 3.27, Epictetus *Diatr.* 4.16.14).[18] But as Abraham Malherbe and others note, the extended use of *siege* warfare as the image of choice is unusual and interesting in its specificity.

Paul depicts a four-stage assault on his opponents: taking up of weapons, tearing down all the ramparts of the enemy stronghold, taking prisoners of war, and bringing his opponents to justice. The general significance of the metaphor is generally unclear, as this dispensation of justice against a rebellious faction is otherwise vague. The allusion to prisoners of war, however, allows us to ascertain Paul's posture toward his opponents. The topic of Roman prisoners of war is a complicated one, further obscured by the fantastical and polemical nature of much discussion from antiquity. Ancient texts give the impression that there were three possible outcomes for Roman prisoners of war: hostages for exchange, enslavement (often leading to participation in gladiatorial games), and execution. Surviving sources are almost exclusively Roman and reflect the biases expected of Romans: they assert how humane Roman prisoner-of-war practices were in comparison to other cultures; claims of Roman humanitarianism are repeated frequently enough that one infers civilian denizens of the Roman Empire were familiar with them.

Enslavement and execution were the most common outcomes for prisoners of war, with the former regarded as an act of clemency. The Roman jurist Florentius, for example, claimed in a folk etymology that "slaves (*servi*) are so called because commanders generally sell the people they capture and therefore save (*servare*) them instead of killing them" (Dig. 1.5.4). Roman writers take pride in their prisoner-of-war policy, contrasting their willingness to let captives live with the Empire's enemies, characterizing the latter as more cruel or brutal—no doubt exaggerating both Roman kindness and foreign malice in the process. We know that family members might be able to purchase the freedom of loved ones; one epigraph records a soldier who purchased his sister's and her children's freedom after she had been enslaved from age 4 to 42 (*SEG* 39.1711). Thus, despite the misery of slavery, preserving one's life with the hope of eventual freedom was preferable to execution.

Paul, however, seems uninterested in mercy for his opponents. Paul depicts his adversaries as a rebellious faction that needs to be brought back to order and punished (*ekdikēsai*). The metaphor of punishment after war against rebels can only refer to execution. The Third Servile War, in which Spartacus led numerous slaves in a revolt against Rome, ultimately resulted in the mass execution of the rebels, a punishment that Roman authors deemed just. In another instance, residents of Illiturgis defected from allegiance with Rome in favor of Carthage; the general Scipio Africanus prefaced the ensuing massacre of Illiturgians by assuring his soldiers that "the time had now arrived when they should take vengeance (*poenas*) for the horrid massacre of their fellow soldiers" (Livy 28.19). Many other examples could be cited. In short, the only "justice" the Roman army would offer a rebellious faction was slaughter. What this meant in practical terms for Paul—that is, what he hoped to do to his opponents at Corinth once he was established as the victor—is unclear. These metaphors indicate that there are severe, but unspecified, "consequences" for not taking him seriously.[19]

Paul uses the image of conquest—albeit not specifically siege warfare—more briefly in his epistle to the Romans (8:37), declaring that Christians will not be conquered by those who wish to separate them from Christ, but in fact are greater than such conquerors. Those in Christ will not be defeated by any war against them, be it from earthly or supernatural powers (8:31–39). Those things that might separate one from Christ—tribulation, distress, persecution, famine, nakedness, danger, and the sword—are stereotypical elements of war in Greco-Roman writings, though Paul seems to use them in two senses: any of these individual components might draw one away from Christ, but taken as a whole they lead him to shift his metaphor to holy warfare in his citation of Ps 44:22 and explicit invocation of conquest in the following verses. Though the metaphorical enemy is different from that of 2 Cor 10:3–6, they both participate in a common discourse of militarily defending one's convictions.

Paul as Soldier of Christ and Prisoner of War—2 Cor 2:14; Phil 2:25; Phlm 2, 23; Rom 16:7

Though Paul occasionally uses military imagery, it is only in his later letters that he explicitly identifies as a metaphorical soldier himself. He twice refers to himself as a soldier of Christ (Phil 2:25, Phlm 2) and twice as a prisoner of war (Phlm 23, Rom

16:7). Both terms occur in the brief letter to Philemon, suggesting that Paul understood these as related concepts, even if he never elaborates on these images. Paul describes his allies in captivity as fellow prisoners of war (cf. Col 4:10), which does little to clarify the specifics of his situation. The word etymologically means something like "fellow captive by spear (*aichmē*)," but designated captives of war more generally and had come to take a metaphorical meaning of "fellow captive under guard" by the classical era. Usage of the term or related words in Greek writings is nonetheless rare and other occurrences do little to elucidate its use by Paul.[20] Its precise significance is unclear, so we can only infer that these people—Junia, Andronicus, and Epaphras—were captive with him at some point and were Christians as well.

Notably, Paul only uses the term in contexts where he claims filiation with others: they are *fellow* soldiers and *fellow* prisoners of war, as he never claims it for himself in isolation. Even when he uses the image less explicitly in 2 Cor 2:14, it nevertheless emphatically applied to "us." Moreover, the terms only occur in greeting sections of his letters (assuming Philippians is a composite document, with a more-or-less complete letter comprising 1:1–3:1). It is difficult to know what to make of this, though the metaphor seems to have been common in religious discourse of the Greco-Roman East.[21] Socrates, for instance, compared his time as a Greek soldier with being a soldier for god: "If, when the commanders whom you chose to command me stationed me . . . I remained where they stationed me, like anybody else, and ran the risk of death. But when the god gave me a station, as I believed and understood, with orders to spend my life in philosophy and in examining myself and others" (Plato *Apol.* 28E). It is likely that Paul draws upon and participates in such traditions, locating himself among the embattled-but-virtuous men of history.[22] Whether or not Paul was illiterate (2 Cor 11:6), these traditions were widespread enough that they were likely "in the air" at the time.

The Praetorian Guard—Phil 1:12–13

Paul claims the "entire praetorium" knows that he is held captive because of Christ. What Paul means with this word is debated, referring to either the edifice termed the praetorium (cf. Mark 15:16; Acts 23:35) or the praetorian guard, ambiguity not helped by Paul's use of a Latin term rather than a more precise Greek word in this instance. The praetorium building could be found in nearly every major city in the Roman Empire and housed governors, kings, and sometimes the emperor himself—the term is also applied to the commander's quarters in Roman military camps. The praetorian guard comprised elite soldiers who served as the emperor's personal army and were based in the Roman capital. If Paul refers to the praetorian guard, then he must be writing from the city of Rome itself; if he refers to a praetorium building, then Paul could be in any number of major cities.

The praetorian guard were probably the most important military force in the empire. The proximity of armed men to the emperor ensured some degree of protection, but also left them vulnerable to the praetorians' whims. The guard held considerable sway over the emperor and directly influenced imperial power: bribery for their allegiance was a major factor in the quick succession of four emperors in 69 CE and five emperors in 193 CE.[23] This is not to mention their frequent role in the assassination of emperors who were either unpopular or otherwise offended them; Nero, Caligula, and many others lost their lives shortly after losing the loyalty of the praetorian guard. These men consequently received numerous benefits—social, financial, etc.—from their imperial benefactors, in hopes that they might not support another imperial claimant. In addition to serving as the imperial bodyguard, the praetorian guard quelled rebellions within the region of Italia and abroad (e.g., Tacitus *Ann.* 2.16, 20.3; Cassius Dio 57.4.3–4), arrested and executed high-level political convicts (e.g., Suetonius *Aug.* 27.4, Tacitus *Ann.* 1.30), were firefighters (e.g., Strabo 5.3.7, Cassius Dio 55.26.4, Appian *Bell. civ.* 5.132), served as security at the annual spectacles (e.g., Tacitus *Ann.* 16.5), collected taxes under Gaius (Suetonius *Cal.* 40), and constructed imperial works (e.g., Josephus *A.J.* 19.257, Suetonius *Nero* 19.2). In chapter 2, we saw that many of these activities were performed by legionaries and auxiliaries in the provinces. The duties of the praetorian guard, by contrast, were limited to Rome's vicinity, except to escort imperial family members on their travels.

If Paul is referring to the praetorian guard in this passage, it is obvious hyperbole. During the Julio-Claudian era, the praetorian guards were comparable to a legion in number—roughly 5,400 men. It is difficult to imagine a scenario where an obscure prisoner would be "known to the whole praetorian guard." Historical evidence for Paul's imprisonment in Rome, however, is scant—it is only attested in a tendentious account in Acts, an account we have seen is laden with heavy symbolism and depends upon elements of Paul's biography that are unsupported by his own letters (e.g., Roman citizenship, rhetorical eloquence). This is not to mention the simple fact that the kind of frequent exchange that the letters to the Philippians presuppose would have been impossible in this fashion between Rome and Philippi.

Some scholars link Paul's imprisonment in Rome with his reference to "Caesar's household" later in the letter (Phil 4:22), understanding this as a clear indication of the imperial household itself. But other interpretations are available. It is also feasible that "Caesar's household" refers to *familia Caesaris*—the high-ranking slaves and freedmen of the emperor who served in a variety of administrative and bureaucratic roles across the empire.[24] These administrators commonly worked in praetoria across the empire, so this reference hardly limits Paul's location to Rome. Indeed, any of Paul's punishments listed in 2 Cor 11:16–33 could have followed a brief detention in or near a provincial or municipal praetorium. Though the praetorium building is more plausible, none of this is conclusive thanks to Paul's ambiguous phrasing.

Conclusion

The authentic letters of Paul are lightly peppered with references to the military, images that Paul uses in an *ad hoc* manner. Paul does not seem particularly invested in any of these military metaphors, nor is he especially familiar with the realia of the military as a social institution. One gets the impression that Paul heard stories of the military and encountered soldiers on occasion, but took no special interest in their lives or combat missions. Consequently, his occasional identification with the military—whether as a soldier of Christ or as a commander attacking his enemies' rhetorical stronghold—do not suggest an endorsement of the Roman Empire or its violence. Military metaphors consistently function as a way of solidifying group identity for Paul: they identify him and his sympathizers as embattled, often in a collective manner.

What might this mean for Paul's politics more broadly? In his earliest letters (1 Thes, 1–2 Cor), Paul anticipates the rescue of Christians from the Day of the Lord. This event involves the destruction and supersession of the Roman imperial way of life in a manner comparable to military conquest. Just as contemporary post-apocalyptic films imagine the end of global capitalism through cataclysmic means, so also does Paul imagine the calamitous end of the Roman way of life. Neither Paul nor such films name the prevailing culture (i.e., capitalism or Rome) directly, but the values these cultures promote and social structures maintaining them are eliminated in one fell swoop. Though the notion of rescue on the Day of the Lord can be found in later epistles (e.g., Phil 1:10, 3:20–4:1), Paul devotes less energy to this scenario as time goes on and the notion of "rescue" from wrath is entirely abandoned. He instead comes to understand the moment wherein Jesus saves an individual to be upon one's death (Phil 1:21–23) and the *mechanics* of salvation occupy his letters to the Galatians and Romans.

"DISPUTED" PAUL

Ephesians

Ephesians is sometimes called the "epistle without a context." It is difficult to ascertain what situation led to the composition of this book and the location where it was produced. Suffice it to say that the writing style is clearly non-Pauline, even if it claims to have been written by the apostle and draws upon his letters as a source of influence. Ephesians articulates a far more developed ecclesiology than was present in any of Paul's letters and more explicitly patriarchal gender norms, suggestive of controversy and ample reflection, perhaps indicating a date sometime around 90–100 CE. As to where it was written, Asia Minor seems as likely as anywhere else in empire, given Paul's authority in that region.

The Armor of God—Ephesians 6:10–17

The authentic letters of Paul used metaphors of weapons and warfare, including a siege against opposing forces (2 Cor 10:3–6), the armor of God (1 Thes 5:8; Rom 13:12), and weapons of one's ministry (2 Cor 6:7). These three images are combined, deployed at much greater length, and are ascribed far more significance in the pseudonymous letter to the Ephesians. Indeed, Christian identity is caught up in militarized identification practices.

Ephesians is brimming with assertions of incomparable peace (1:2, 2:14, 2:17, 4:2) and love (2:4, 3:17, 4:2, 4:15–16, 4:26, 4:32, 5:2, 5:28, 5:33, 6:23–24), so it is somewhat surprising that its Christian body-building (*eis oikodomēn tou sōmatos tou Christou*; 4:12) is an overtly martial process. Christ and God are explicitly attributed a kingdom in the heavenly places (5:5).[25] The author pays homage to God through standard demonstrations of subservience before the head of state (3:14) and claims to be his ambassador to the nations (6:20). While God's realm is primarily in heavenly places, his authority also extends to the earthly realm where his vice-regent Christ intervenes as necessary (1:20–23).[26] Notable among Christ's imperial deeds (*res gestae*) is the establishment of peace between hostile neighbors: Jews and Gentiles (2:11–22). Christ accomplished this reconciliation by expanding God's territory and annexing an adjacent population. To acquire this larger territory, Christ destroyed the defensive partition separating the two domains that he might grant his new Gentile citizens the benefits of Israel's governance (*politeia*).[27]

While the universal church comprises the body of the cosmic Christ throughout Ephesians, God is attributed a much more anthropomorphic figure. God's right hand was noted early in the letter (1:20), but a fuller image of the divine body comes forth in the lengthy description of the deity's armor in Eph 6:10–17: a belt for his waist, a breastplate, shoes for feet, a shield to protect the abdomen, a helmet for the head, and finally a sword for that same right hand. Christians should equip themselves with this gear to prepare for holy combat. But rather than rushing into battle, Christians should opt for a defensive stance against their opponents—they must stand firm against the devil and the cosmic powers on the day of evil.

Some commentators read Ephesians as subverting Roman military norms: Ephesians was not the product of sympathy to Rome, but an author critical of Roman violence.[28] These anti-imperial interpretations contend that Ephesians used military metaphors to undermine Roman hegemony by imagining a deity exceeding all earthly authorities (1:21) and spiritualizing combat as a criticism of Rome's earthly military might (6:12). There are important reasons to doubt an anti-Roman reading of Ephesians, but even if this approach accurately characterizes the social program the author tried to advance, this reading would fail to appreciate how crucial the military is to the epistle's imaginary. The centrality of the military becomes clear when one considers the difficulty of a parodic understanding of Ephesians' metaphor of political annexation: Ephesians uses a prolonged metaphor of Christ's unification of Jews and Gentiles, bringing the latter into the former.

Two aspects of Christ's role as military vice-regent vis-à-vis territorial expansion might elucidate the problem. First, there are parallels to the ostensive benevolence of Roman peace through inclusion into the empire, whether by client kingship or annexation. Philip Freeman demonstrates that the distinction between military conquest and peaceful annexation was entirely retrospective—it was not obvious the extent to which the process would be accepted as legitimate until the dust had settled.[29] For instance, the kingdom of Nabataea was invaded and reduced to provincial status under the name Arabia in 106 CE. Rome did not offer an official characterization of this annexation until 111 CE, at which point coinage insisted it was acquired (*Arabia adquisita*) and not captured (*Arabia capta*). This alone should lead one to be hesitant to accept assertions of compassionate territorial expansion. Ephesians explicitly identifies Christ as the agent of the benevolent destruction that culminates in a territorial and political unity of Jews and Gentiles, long-hostile neighbors. It may be tempting to explain this as irony-laden appropriation of Roman language that nevertheless rejects imperial structuring of state relations. Such a position does not account for how Roman valuation of these metaphors are reiterated in Ephesians, including the fascination with the power constitutive of God's and Christ's cosmic sovereignty as well as its implications for practices of spiritual warfare. The desire to extend one's political reach across the world and the violence to achieve it remain unquestioned in Ephesians.[30]

Second, and more crucially, is that the validity of military violence is woven into the fabric of Ephesians' argument. Imperialism is not limited to a few isolated passages that could be easily separated from the epistle's themes, logic, and explicitly normative understanding of Christian identity. The military played a significant role in the development of early Christian ideology; Ephesians' author could not "think outside the imperial box."[31] That is, even if imperial terminology *really* served to discuss Christ, the author made recourse to army-talk to articulate those thoughts. The medium of expression, namely language specific to Roman state violence, should not be dismissed as peripheral to Ephesians' politics. Whatever the intention of the author of Ephesians may have been, the effect would be almost entirely congenial to Roman state interests.

Ephesians may be characterized as a "Roman" document not only in that it was composed under the Roman Empire, but that it advanced interests favorable to the Roman state—regardless of the author's intent. The epistle's efforts to incorporate Christians into typical Roman habits of living and family structures, as well as their recognition of certain types of activity as laudatory and expression of such laudation, suggest that any implied criticism of Roman imperialism in Ephesians is severely truncated. Like many pseudo-Pauline writings, Ephesians was interested in ensuring Christianity's adherence to Roman respectability politics.

Pastoral Epistles

Even more than Ephesians, the Pastoral Epistles indicate anxiety about gender norms and ecclesial hierarchy within the early church. The authentic Pauline letters are generally flippant in disregarding the patriarchal household (e.g., 1 Cor 7:8, 7:25–28) and seem to use church titles more or less interchangeably; by contrast, the Pastoral Epistles weaken women's authority whenever possible and not only articulate a clear ecclesiology, but even require qualifications for church offices.[32] This is not to mention the use of a literary style that differs significantly from the authentic letters. The Pastoral Epistles evince major concerns with Greco-Roman "respectabil-ity politics," that is, adhering "respectable" social norms of the Roman Empire and being model citizens; the authentic epistles' lack of interest in these matters (see, e.g., the apostle's confusing position on divorce in 1 Cor 7, boasting about his shameful experiences in 2 Cor 11:16–33) were tamed and brought into the orbit of prevailing Roman discourse. It is generally thought that the Pastoral Epistles were composed by the same author (though this is disputed as well), probably written somewhere in the Roman East 90–110 CE in an urban context.

Soldiers of Christ—1 Tim 1:18; 2 Tim 2:3–4

While Paul's military metaphors mostly describe either cosmic battles or his own precarious situation, the pastor deploys them to illustrate Christian values more broadly. In 1 Tim 1:18, for instance, the author commands "Timothy" to battle false teachers. This wording places the pseudonymous "Paul" in a position of military authority and identifies "Timothy" as a soldier. Whereas authentic Paul's military metaphors are laden with irony for his own situation, the pastor not only uses the military with a serious intonation, but does so in a manner leading into an admoni-tion to pray for "kings and others in authority" (1 Tim 2:2). That is, the image of the soldier is not merely an ad hoc metaphor to clarify the author's meaning, but corresponds to clear practices within their social world, practices entailing respect for the Roman "state" and its representatives. While the authentic Paul had little interest in identifying with the empire, the pastor took the validity of Rome and its conven-tional values as a starting point for developing his own practical ethics.

In the Second Letter to Timothy, the pastor compares Christian values to the work ethic of soldiers, athletes, and farmers (2:3–4). The pastor admires soldiers' willingness to suffer and how they prioritize their commanders' happiness over their own affairs. The first of these, readiness to suffer, is well attested as a military value in antiquity. Tacitus, with more than a touch of hyperbole, writes, "even in a beaten army when every tenth man is felled by the club, the lot falls also on the brave" (*Ann.* 14.44); Seneca offers similar thoughts, "the perfect man . . . believed that he was citizen and soldier of the universe, accepting his tasks as if they were his orders. Whatever happened, he did not spurn it as if it were evil and borne in upon him by hazard; he accepted it as if it were assigned to be his duty" (*Ep.* 120.12). There was a similar literary trope concerning the death of the cowardly soldier: he died face-down, fleeing from battle, rather than the brave soldier who died face-up from battle wounds. Thus, the Roman rhetorician Aelian recounts the words of a man facing certain death in battle, "Do not inflict a shameful and cowardly wound, but strike me in the front of the chest, lest my beloved convict me of cowardice and be wary of laying out my corpse." (*Nat. an.* 4.1) This idealization of soldier's bravery in the face of danger is a common trope in ancient literature.

The obedience of soldiers to their commanders was a major concern, too, and there is evidence that most soldiers were deeply loyal to their generals. It was not uncommon for auxiliaries to adopt their commander's name as part of the *tria nomina* when they received their citizenship. Commanders fostered this loyalty in many ways, most often through the Roman *disciplina militaris*, a cluster of values that sought to impose elite values of self-control on the rabble they believed to comprise the army's ranks.[33] This specific variety of military discipline had some important features: a) limiting pleasures available to soldiers (e.g., legionaries' marriage ban, Hadrian's ban on gardens within camps), b) mandatory manual labor to prevent idleness (e.g., construction, paving, digging trenches), c) rare instances of great reward for excellent service (e.g., donatives, promotions), and d) commander's visible participation in these austere conditions. This last element was the most significant. The emperor Hadrian, for example, purportedly ate soldiers' gruel during military campaigns and walked twenty miles in battle armor (HA *Hadr.* 10.2, 10.4). This relationship facilitated officers' and emperors' patronage of their soldiers, allowing them to be the primary conduits of life's pleasures. Such an approach was not always successful, leading to mutinies and discontent on occasion, but these were rare. This policy was built upon elite stereotypes of the lower classes; only by cultivating elite habits could the *ferocia* of the lower-class mob of the army's soldiers be tamed.

It is clear that the pastor has a better understanding of the military's workings and values than Paul. This, perhaps unsurprisingly, correlates with the pastor's espousing a far more favorable attitude toward Rome and its military than Paul. Like author of Ephesians, the pastor adopts these metaphors for his own purposes, but they do not have an ironic function as they tend to in the authentic letters of Paul.

NOTES

1. E.g., Peter W. Macky, *St. Paul's Cosmic War Myth: A Military Version of the Gospel*, Westminster College Library of Biblical Symbolism 2 (Frankfurt: Lang, 1998).

2. Friesen, "Poverty in Pauline Studies"; Friesen, "Paul and Economics."

3. Pervo, *Dating Acts*, 51–147; Ryan S. Schellenberg, "The First Pauline Chronologist? Paul's Itinerary in the Letters and in Acts," *JBL* 134 (2015): 193–213.

4. See, e.g., Gregory P. Fewster, "'Can I Have Your Autograph?': On Thinking about Pauline Authorship and Pseudepigraphy," *Bulletin for the Study of Religion* 43/3 (2014): 30–39.

5. See, e.g., Isaac, *Limits of Empire*; Millar, *Roman Near East*; Sartre, *Middle East under Rome*; Sherwin-White, *Roman Foreign Policy*.

6. See the influential discussions in Helmut Koester, "Imperial Ideology and Paul's Eschatology in 1 Thessalonians," in *Paul and Empire: Religion and Power in Roman Imperial Society*, ed. Richard A. Horsley (Harrisburg: Trinity Press International, 1997), 158–166; John Dominic Crossan and Jonathan L. Reed, *In Search of Paul: How Jesus's Apostle Opposed Rome's Empire with God's Kingdom* (San Francisco: HarperSanFrancisco, 2004), 167–177.

7. See the discussion of the phrase and Paul's critical use of it in Klaus Wengst, *Pax Romana and the Peace of Jesus Christ*, trans. John Bowden (London: SCM Press, 1987), 19–21, 37–38, 77–78.

8. See the classic, if overstated, discussion of this other usage in G. D. Kilpatrick, "Acts vii.52 ΕΛΕΥΣΙΣ," *Journal of Theological Studies* 46 (1945): 136–145.

9. David R. Catchpole, "The 'Triumphal' Entry," in *Jesus and the Politics of His Day*, ed. Ernst Bammel and C. F. D. Moule (Cambridge: Cambridge University Press, 1984), 319–334. The specific elements of this schema are debated; see the following footnote for alternatives.

10. Catchpole, "Triumphal"; Paul Brooks Duff, "The March of the Divine Warrior and the Advent of the Greco-Roman King: Mark's Account of Jesus' Entry into Jerusalem," *JBL* 111 (1992): 55–71; Adela Yarbro Collins, *Mark: A Commentary*, Hermeneia (Minneapolis: Fortress, 2007), 514–521; Leander, *Discourses of Empire*, 255–267.

11. Neil Elliott, "The Apostle Paul and Empire," in *In the Shadow of the Empire: Reclaiming the Bible as a History of Faithful Resistance*, ed. Richard A. Horsley (Louisville: Westminster John Knox, 2008), 105, however, sees 1 Thes 4:12 as advocating a more positive stance to outsiders.

12. Neil Elliott, "The Apostle Paul's Self-Presentation as Anti-Imperial Performance," in *Paul and the Roman Imperial Order*, ed. Richard A. Horsley (Harrisburg: Trinity Press International, 2004), 77. See, with some important modifications needed, Cilliers Breytenbach, "Paul's Proclamation and God's 'thriambos': Notes on 2 Corinthians 2:14–16b," *Neotestamentica* 24 (1990): 257–271. See also the incredibly detailed study Christopher Heilig, *Paul's Triumph: Reassessing 2 Corinthians 2:14 in Its Literary and Historical Context*, Biblical Tools and Studies 27 (Leuven: Peeters, 2017). Cf. the word's use in the pseudo-Pauline Col 2:15.

13. Thomas R. Yoder Neufeld, *"Put on the Armour of God": The Divine Warrior from Isaiah to Ephesians*, JSNTSup 140 (Sheffield: Sheffield University Press, 1997), 73–93 presents influential arguments to the contrary; he sees 1 Thes 5 as sustained anti Roman polemic.

14. See Konstantin Nossov, *Gladiator: The Complete Guide to Ancient Rome's Bloody Fighters* (Guilford: Lyons, 2011), 57–58, 92; cf. *CIL* 13.1997; Artemidorus *Onir.* 33.

15. *Contra* Lisa M. Bowens, "Investigating the Apocalyptic Texture of Paul's Martial Imagery in 2 Corinthians 4–6," *JSNT* 39 (2016): 3–15, which argues for a continuous martial metaphor throughout 2 Cor 4–6. I would suggest the military images are incidental, given the variety of metaphors the apostle uses throughout these chapters, with little regard for consistency even over the course of a few verses.

16. For persuasive arguments that Paul writes concerning soldiers' food and not their wages, see Chrys C. Caragounis, "ΟΨΩΝΙΟΝ: A Reconsideration of Its Meaning," *Novum Testamentum* 16 (1974): 51–52.

17. *CPL* 304, translation from Davies, *Service in the Roman Army*, 201–202.

18. Abraham J. Malherbe, "Antisthenes and Odysseus, and Paul at War," *Harvard Theological Review* 76 (1983): 143–173; George H. Guthrie, *2 Corinthians*, Baker Exegetical Commentary on the New Testament (Grand Rapids: Baker Academic, 2015), 472–475.

19. The image of military assault is further developed later in the letter. Though obscured by most English translations, Paul uses a similar metaphor in 2 Cor 11:12: the phrase often rendered something like "undermine the claim" (English Standard Version) or "deny an opportunity" (New Revised Standard Version) literally means "undercut the base of operations" (*ekkopsō tēn aphormēn*).

20. Wansink, *Chained in Christ*, 170, esp. n. 76; Gerhard Kittel, "αἰχμάλωτος, αἰχμαλωτίζω, αἰχμαλωτεύω, αἰχμαλωσία, συναιχμάλωτός," *TDNT* 1.195–197. They cite Rom 7:23; 2 Cor 10:5; Luke 21:24; Demades *Orat.* 1.65; Sextus Empiricus *Math.* 1.295; Pseudo-Lucian *Ass* 27.

21. Franz Cumont, *Oriental Religions in Roman Paganism*, 2nd ed. (New York: Dover, 1956), 213 n. 6, citing Apuleius *Metam.* 11.14; Vettius Valens 5.2, 7.3; Minucius Felix 36; Propertius 1.6.30, 4.1.137; Horace *Odes* 3.26; Ovid *Amor.* 1.9, *Ars amat.* 3.233; Plato *Apol.* 28E; Seneca *Ep.* 107.9. Cf. Epictetus *Diatr.* 3.22.69.

22. See additional arguments to this effect in Wansink, *Chained in Christ*, 147–174.

23. On the praetorian guard, see Sandra Bingham, *The Praetorian Guard: A History of Rome's Elite Special Forces* (London: Tauris, 2013); Guy de la Bédoyère, *Praetorian: The Rise and Fall of Rome's Imperial Bodyguard* (New Haven: Yale University Press, 2017).

24. See the extensive analysis of *familia Caesaris* as an institution and their place in early Christianity in Michael A. Flexsenhar, III, *Christians in Caesar's Household: The Emperor's Slaves in the Makings of Early Christianity*, Inventing Christiantity (University Park: Pennsylvania State University Press, 2019). For compelling arguments to the contrary, see Angela Standhartinger, "Letter from Prison as Hidden Transcript: What It Tells Us about the People at Philippi," in *The People beside Paul: The Philippian Assembly and History from Below*, ed. Joseph A. Marchal, SBLECL 17 (Atlanta: SBL, 2015), 129–130 n. 90.

25. This discussion draws upon Christopher B. Zeichmann, "Liberal Hermeneutics of the Spectacular in the Study of the New Testament and the Roman Empire," *Method and Theory in the Study of Religion* 30 (2018): forthcoming.

26. On Christ as God's vice-regent in Ephesians, see Julien Smith, *Christ the Ideal King*, WUNT II 313 (Tübingen: Mohr Siebeck, 2011), 175–206.

27. Harry O. Maier, *Picturing Paul in Empire: Imperial Image, Text and Persuasion in Colossians, Ephesians and the Pastoral Epistles* (London: Bloomsbury, 2013), 103–136 instead understands Ephesians' primary metaphor as civic, particularly with reference to rival cities that require a righting of relations via *homonoia*; though Maier's reading is compelling, it has difficulty accounting for the wall and military imagery throughout the letter.

28. E.g., Timothy G. Gombis, *The Drama of Ephesians: Participating in the Triumph of God* (Downers Grove: InterVarsity, 2010), 155–179; N. T. Wright, "Paul and Empire," in *The Blackwell Companion to Paul*, ed. Stephen Westerholm (Oxford: Wiley-Blackwell, 2011), 285–297; Fredrick J. Long, "Ephesians: Paul's Political Theology in Greco-Roman Political Context," in *Christian Origins and Classical Culture: Social and Literary Contexts for the New Testament*, ed. Stanley E. Porter and Andrew W. Pitts, Texts and Editions for New Testament Study 9 (Leiden: Brill, 2013), 255–309.

29. Philip Freeman, "On the Annexation of Provinces to the Roman Empire," *Classics Ireland* 5 (1998): 30–47. Likewise, one aureus minted in September 70 CE declares *Iudaea recepta*, a message soon revised in favor of the more famous *Iudaea capta*.

30. The confluence of Christ's cosmic headship (with all its connotations of power, autonomy, and might) with the household code's laudation of subjection, discipline, and mastery is also normative of the Christian *paterfamilias* (5:21–6:9). Christ's similar functions as head of the church could be understood as Ephesians' extension of the emperor's function as *pater patriae* to Christ.

31. Arthur J. Droge, "Ghostlier Demarcations: The "Gospels" of Augustus and Mark," *Early Christianity* 2 (2011): 343, though discussing the Gospel of Mark.

32. Dennis R. MacDonald, *The Legend and the Apostle: The Battle for Paul in Story and Canon* (Louisville: Westminster John Knox, 1983) presents a compelling scenario for the Pastoral Epistles' anxiety over gender norms, though he overstates evidence that the Pastoral Epistles made use of the *Acts of Paul and Thecla*.

33. See Phang, *Roman Military Service*, 359–378; Clifford Ando, "The Army and the Urban Elite: A Competition for Power," in *A Companion to the Roman Army*, ed. Paul Erdkamp, BCAW (Malden: Blackwell, 2007), 359–378.

5

The Military in Revelation

Leif Vaage aptly characterizes Revelation as "prone to excessive interpretation."[1] This tendency is understandable, given how opaque the text is. Interpreters, both academic and otherwise, have consequently proposed various *keys* to Revelation's interpretation, the most frequent of which in recent scholarship has been the author's contempt for the Roman Empire. Passages that were formerly interpreted as allegories about God and evil, or insights to an anticipated Day of the Lord, are now generally seen as the product of an author fantasizing about the destruction of Rome. Though much of the book's significance remains disputed, the scholarly consensus on this matter positions us for assessing the author's posture toward the Roman military.

Revelation was composed by John on Patmos, an island off the western coast of Asia Minor (1:9). Commentators both ancient and modern have identified John of Patmos with the apostle John and John the elder mentioned by the church father Papias, though there is little evidence to support either contention. More likely, John of Patmos is an otherwise-unknown figure within early Christianity. Interpreters have long agreed that the author was writing during the later portion of the emperor Domitian's reign (81–96 CE).[2] At the very least, Revelation was composed after 70 CE—evident in the use of "Babylon" (which destroyed the first Jerusalem temple) as a cipher for Rome, the legend of Nero's postmortem revival in Rev 13:3 (having died in 68 CE), and the imagery of Rome's seven hills on a coin of Vespasian (Rev 17:9; beginning his reign 69 CE). Even with the "key" of anti-Roman polemic, we are still in very murky waters with Revelation. What might the study of the Roman military offer interpreters of Revelation?

Two-Edged Weapon from the Mouth—Rev 1:16; 2:12; 2:16; 19:15; 19:21; cf. 6:8

New Testament authors usually use the standard Greek word *machaira* to refer to swords (e.g., Matt 26:52, Eph 6:17, Heb 4:12), but Revelation prefers a more obscure word: *rhomphaia*. Ancient literary texts indicate that it is a Thracian weapon (Livy 31.39, Plutarch *Aemil.* 18.5), but it is unclear whether this word refers to a sword, a spear, or some other type of weapon. The obscurity of this word is evident in the confusion of biblical translators of antiquity: the Vulgate translates *rhomphaia* as a Germanic spear (*framea*; Ps 9:6) and the Syriac version renders the word as "lance" (Luke 2:35, Rev 1:16). Fortunately, archaeological discoveries have clarified the word's meaning to some extent. Christopher Webber describes the *rhomphaia* as "a two-handed cutting weapon with a long handle and a long, straight or slightly curved single-edged blade."[3] Its blade were usually 2.5 cm wide and 50–60 cm long, having a hilt of somewhat shorter length, totaling about a meter in length. Surviving *rhomphaiai* date to the Hellenistic era, suggesting it went out of use due to the Romanization of the Thracian royal army during the first century BCE.

The identification of the *rhomphaia*, however, only complicates matters. The weapon is consistently described as "two-edged" in Revelation, which the *rhomphaia* was not. But even granting this artistic license, what are we to make of the strange portrait of a large, sharp, and double-edged weapon protruding from the mouth of the glorified Christ? Revelation uses this image to evoke Isa 49:2 and perhaps Wis 18:15–16 as well. In the Septuagint version of the former, God makes his servant's mouth like a sharp sword (*machaira*) and in the latter God's word is compared to a warrior with a sharp sword. Commentators are divided over whether the image in Revelation indicates the sharpness of the prophetic tongue, or "the irresistible power of divine judgment."[4] The matter may be clarified upon consideration of the physical features of the *rhomphaia*: namely that John is conflating distinct images, further alluding to the "rod of [God's] mouth" which will strike the earth in Isa 11:4 along with the two aforementioned references. Though John has famously poor Greek, misspelling words and using poor grammar throughout Revelation, it is clear he knows that *machaira* is the standard word for sword (6:10, 13:10, 13:14). Writing a reasonable sailing distance from Thracia—and with Thracian auxiliaries posted throughout the Levant—John's passing familiarity with the *rhomphaia* may have led him to treat it as a bladed weapon (so Isa 49:2 and Wis 18:15–16) that is also a rod (so Isa 11:4). This finds validation in Rev 19:15, where Isa 11:4 is explicitly cited in the context of Christ's *rhomphaia*. The *rhomphaia* allowed John to have the best of both intertextual worlds, evoking multiple passages which indicate the deadly consequences of wickedness.

THE LAMB OPENS THE FIRST SIX SEALS—REV 6:1–17

Sealed scrolls are a symbol of God's plan in Revelation; the Lamb opens them to unleash the calamitous effects on the world, the first four of which are denoted by ominous horsemen. Military imagery abounds in this chapter of Revelation, especially its description of the first two riders. The most plausible interpretation contends that these riders represent military conflict outside Rome's borders (the first rider, on the white horse) and military conflict within Rome's borders (the second rider, on the red horse). The first rider has a bow; archers, especially mounted ones, were associated with Rome's eastern enemies, such as the Parthians, Sarmatians, and Getans.[5] Though there were many Roman auxiliary units composed of archers, including mounted archers, *rhetorically* and *literarily* the bow signified Rome's enemies. For instance, it was almost exclusively auxiliaries recruited from the Roman East that were archers: Thracians, Syrians, Cretans, etc.[6] This was not a standard "Roman" weapon and, even within the ranks of the Roman army, mostly noncitizen easterners used a composite bow. Moreover, this rider comes as a "conqueror in order that he might conquer"—language best characterizing a foreign threat. Craig Koester is probably correct that this image does not evoke the Parthians in particular, but draws upon a vague image of a foreign threat from the east; writing from the Roman East, that image tended to look rather Parthian in the popular consciousness.

The second rider, by contrast, can remove the existing peace from earth, leading people to murder one another. This probably refers to the end of *pax Romana* within Rome's borders and the chaos would ensue. Unlike the "conquest" of the first rider, this rider simply allows people to "slaughter" each other. This rider also holds a large sword (*machaira*) and the red color of his horse is appropriate for the bloodshed he brings. Some scholars interpret this violence as an outbreak of brigandage, but this is unnecessary: John implies that murder occurs among peers. This could mean revolts, neighbor-against-neighbor scenarios, or something along these lines. Brigands, despite the insistence of those working on the phenomenon of "social banditry," were almost always perceived negatively by Roman civilians; as discussed in chapter 2, they were not perceived as peers, but as predators lurking and preparing for their opportunity to take advantage of everyday folk. Brigands were barely human in imagination of the Roman East, and certainly not the "Robin Hood" figures they are sometimes made out to be.

The third beast, Famine, says little of interest here, but the fourth rider, Death with the consort Hades, kills a quarter of the world population, by means that included the sword (*rhomphaia*). It is difficult to ascertain why John used the word *rhomphaia* in this instance, though it may be an accident where *machaira* would be preferred. The use of a weapon presumably refers back to the second rider's sword, as each variety of death in this passage corresponds to a given horse-rider. The fifth seal likewise says little of interest, as it is unclear who kills the martyrs.

The sixth seal leads kings, tribunes, and great men, among others, to hide in the mountains. While the previous five seals were satanic and had negative consequences, the sixth is representative of Revelation's revenge fantasy against those "persecuting" the Christians of Asia Minor. This fantasy understands the persecutors as the Roman Empire, here synechdotally represented by its most powerful. The reference to "the kings of the earth" presumably not only includes client kings, but the emperor himself, despite how emperors persistently distanced themselves from this title (see the discussion on page 96 above). John of Patmos apparently used this formulation not out of spite for imperial rejection of this title, but to evoke the nearly identical phrasing of Isa 34:12: "kings, rulers, and great ones." G. K. Beale notes, "as in Isaiah 34, so here these groups of people undergo divine judgment because they are an essential part of the corrupt world system, which must be destroyed."[7] That is, few kings were around the empire by the time of Domitian's reign, but John lends credibility to his vision of judgment by couching it in the archaic and authoritative language of Isaiah.

SOLDIER-LIKE LOCUSTS—REV 9:7–11

When the fifth angel sounds his trumpet, locusts attack people who lack the mark of God. The description of these locusts initially seems very peculiar, as they resemble horses, kings, and humans prepared for battle, among other creatures: "The locusts looked like horses prepared for battle. On their heads they wore something like crowns of gold, and their faces resembled human faces. Their hair was like women's hair, and their teeth were like lions' teeth. They had breastplates like breastplates of iron, and the sound of their wings was like the thundering of many horses and chariots rushing into battle." This draws upon the scenery of Joel 1–2, wherein a plague of locusts commences with the sound of the trumpet and resembles an invading army: the locusts are "like the appearance of horses, and like war horses so they run. With a noise like that of chariots they leap over the mountaintops, like a crackling fire consuming stubble, like a mighty army drawn up for battle. At the sight of them, nations are in anguish; every face turns pale. They charge like warriors; they scale walls like soldiers. They all march in line, not swerving from their course" (Joel 2:4–7; cf. 1:16). Much of Revelation's description of the locusts is taken directly from Joel, evoking a familiar image and borrowing authority from the Hebrew Bible.

Revelation's use of Joel allows us to ascertain Revelation's distinctive voice where these parallels are absent. Joel lacks parallels to the women's hair, or for our interests, the iron breastplates or crowns of gold. These are both generic images attested in ancient literature across cultures. A number of interpretations have been proffered; some see this as indicating a demon-like quality to the locusts, others that the locusts were rational beings, still others as evoking the Hebrew Bible (e.g., Job 39:19–25), and Bruce Malina and John Pilch even suggest that Revelation depicts the locusts as centaur-like creatures comprised of different animal parts.[8] These, especially the final suggestion, seem to overlook Revelation's dream-like approach to its images, taking them too literally. The shifting metaphors and imagery are typical of Revelation, evoking a series of animal-military images that do not require the internal consistency that Malina and Pilch suggest. Consider, for example, how pure linen and scarlet clothing is representative both of evil forces (18:16) and heavenly forces (19:13–14); while some would argue that Revelation's apparent inconsistency is best explained as demonic imitation of the divine, there is no indication John understands its images this way. Rather, Revelation positions the locusts as a foreign and bizarre threat to those who oppose the Lord, the specific appearance of which does not require consistency. One need only know that they will inflict as much suffering (albeit without death) as an army.

MARAUDING ANGELS—REV 9:13–21

The sixth angel's trumpet portends the release of four destructive angels located at the Roman Empire's eastern limit of the Euphrates as well as an incalculably large cavalry force to destroy one third of humanity. Though they recall a different group of four angels mentioned in 7:1, the earlier angels *prevented* destruction and were located at the four corners of the earth, rather than were forces of destruction restrained at a single corner of the world. As elsewhere, Revelation reuses imagery in a manner different from its initial appearance. Such a change makes sense in this context, as the image of four angels of destruction was commonplace in Jewish apocalyptic literature (cf. 1 En. 56:5–8; 2 Bar. 6). The angels are also familiar in that destructive forces are represented as eastern threats to the Roman Empire, similar to the Parthian-like forces released by the opening of the first seal.

The angels are accompanied by an innumerably large cavalry force consisting of horses breathing fire, smoke, and sulfur, with breastplates of corresponding color: red, blue, and yellow. Their number is a matter of contention among scholars, being either 20,000 or 200,000,000.[9] The exact number is inconsequential, as it indicates a number that is incalculable in most contexts, but known with precision to John (9:16). But the riders themselves bear no real responsibility for the destruction, as their bestial steeds are the ones who inflict injury with their tails and plagues of fire, smoke, and sulfur. The horses themselves are described in a grotesque manner like the locusts brought about by the preceding trumpet: they have heads like lions and tails like snakes. Again, this should not be interpreted literally.[10] The point, as with the number of cavalrymen themselves, is that they cannot be understood by most, but John alone experiences their fantastic nature and relates it with as much accuracy as possible to the reader.

MICHAEL BATTLES THE DRAGON—REV 12:7–8

The archangel Michael is responsible for defeating and exiling the dragon, identified as the serpent of Eden, the devil, and Satan—the latter two being synonymous in Hebrew and Greek. Michael's angelic army appears only once in Revelation and is never mentioned in other Jewish literature. John imagines heavenly warfare similar to that which occurs regularly in the earthly realm. The notions of spiritual warfare are not specifically Roman, but Revelation understands the military as a means of using power-by-proxy and a force in competition for disparate peoples. While that may not strike the modern reader as particularly novel, this reflects political and military developments of the Hellenistic and Roman periods, developments in turn reflected in theological discourse as well. New Testament spiritual warfare usually depicts God as commander of the heavenly army, though his subordinates often perform the most visible functions of this military. Satan likewise has an army of angels at his command.

The spiritual warfare of the Hebrew Bible, by contrast, is interested in the Lord's *personal* and *direct* participation in military conquest. This military role is consistent with the Babylonian-Achaemenid context in which these texts were composed: national warrior gods aided their people in battle, just as the king himself went to battle alongside his warrior-citizens in terrestrial combat. As this mode of warfare changed, so also did this archetype of deities lose popularity: political power was increasingly experienced as something distant with the rise of international empires. The archetype of the national warrior god was no longer in widespread use during the Hellenistic and Roman periods, but the military inflection of the Jewish deity remained significant in Jewish and Christian literature. Thus, despite the absence of much specifically *Roman* about the cosmic battle depicted in Revelation (e.g., Roman engineering, Roman combat tactics), this battle is nevertheless helpfully understood as such. Judaism was nudged from a henotheistic theology toward a more clearly monotheistic worldview that led to a clearer dualism in its moral and cosmological frameworks. Consequently, the warrior god of Judaism no longer battled deities of neighboring states, but an unholy kingdom ruled by his only real foe: Satan. For Revelation, the Lord sits at a considerable distance from the terrestrial realm of his subjects, where his delegate Jesus battles until an appointed time. This resembles the modes of power and authority that prevailed in the Roman period, where the emperor remained mostly inaccessible, but was experienced by means of his proxies and subordinates. Moreover, the goal of divine conquest shifted from territorial expansion to absolute terrestrial authority: no longer were Edom, Canaan, and Philistia the desired dominions of the Lord, but the inhabited whole world. As Luther Martin has shown, this is typical of Hellenistic cosmologies in general and may be attributed toward a shift in conceptions of political and military power attending to the rise of imperial administrations.[11]

CONQUERING BEAST—REV 13:7–8

The image of the conquering beast unambiguously elicits that of the emperor: he has power over all, regardless of clan, language, or ethnicity. The beast is a figure with unparalleled hubris, demanding that all worship him, blaspheming the Lord, and "making war" upon the saints. Many interpreters take this as an allegory concerning one or another emperor—most often Domitian—with a purportedly violent policy against Christians. This fairly literal interpretation, however, is difficult to sustain. There is no evidence that Domitian (or other living emperors) demanded worship or claimed to be a living God, for instance;[12] likewise, evidence of Christian persecution during the first century is often overstated.[13]

The beast's "war" upon the saints should probably not be interpreted literally or even metaphorically. Rather, it seems akin to Paul's self-identification as a prisoner of war—that is to say, a provocative exaggeration that attempts to unite a disparate Christian group around the notion of a shared trauma. There is a long history within Christianity of conflating historical martyrdom (e.g., Jesus, John the Baptist) with fictionalized-and-fantastical martyrdoms (e.g., Eusebius' martyrs, probably much of Acts). Likewise, there is a propensity to exaggerate any Christians in conflict with the state (ranging, e.g., from Paul's brutal punishments to discourse on taxation), often deploying language of warfare to describe any perceived persecution, however frivolous it might be—one need only recall present discourse about the "war on Christmas" or the "politically correct war on Christianity." There is an understandable appeal in identifying as part of a righteous-but-persecuted minority. Indeed, scholars have recognized for decades that Revelation's language of persecution is more rhetoric than reality. Thus, Adela Yarbo Collins contends that "*relative*, not absolute or objective, deprivation is a common precondition of millenarian movements. In other words, the crucial element is not so much whether one is actually oppressed as whether one *feels* oppressed."[14] Likewise, Leonard Thompson contends that Revelation derives from "John's perspective on Roman society rather than significant hostilities in the social environment."[15] What exactly was the "war" that John wrote concerning? There is no way of knowing for certain, but suggestions include the expansion of the "Jewish tax" to Christians under Domitian,[16] controversies within synagogues, among many other possibilities.

TEN KINGS AND THE FIRST BATTLE—REV 17:15–18:24

Through the not-very-subtle cypher of "Babylon," John describes a vivid fantasy about the destruction of Rome. The woman—in a depiction permeated with misogyny—is the "great city that rules over the kings of the earth" (17:18), which can only refer to the capital of the empire. The woman sits upon seven hills, which further identifies her with Rome; there are numerous literary references to Rome's seven hills: Vergil *Aen.* 6.782–783, *Georg.* 2.535; Martial *Epig.* 6.64; Cicero *Att.* 6.5, among many others. But more important for our purposes is a coin depicting a woman personifying Rome sitting upon seven hills, holding a sword, was minted in 71 CE in the province of Asia—the province containing all the churches with whom John corresponds.[17] The image was thus well known not only in the city of Rome, but in Asia Minor as well, and David Aune makes a compelling argument that John drew upon this image as a means of criticizing Rome. The identity of the ten kings who first act as consorts for the woman before bringing about her death is unclear. These could be the kings of the east mentioned in 16:12, comprising a coalition that threatens Rome.[18] These kings should be distinguished from the kings that consort with the woman and lament her demise (18:3, 9–10).

The violent fantasy of Rome's destruction at the hands of the kings uses a number of war-related metaphors. First, the sexual assault of the woman (17:16) evokes a common wartime practice. As discussed in the context of Matthew's Healing of the Centurion's Slave earlier, it was not uncommon for Greek and Roman soldiers to kidnap and rape youth and women during the process of domination and conquest, the brutal process of which Revelation describes here. This image was both vivid and common in antiquity. A sculpture from Aphrodisias depicts Claudius capturing a personified Britannia implying rape and many authors describe the sexual brutality of the Roman military during combat (see Figure 3.3). Consider Tacitus' description of the sack of Cremona in 69 CE:

> Forty thousand armed men burst into the town; the number of camp-followers and servants was even greater, and they were more ready to indulge in lust and cruelty. Neither rank nor years protected anyone; their assailants debauched and killed without distinction. Aged men and women near the end of life, though despised as booty, were dragged off to be the soldiers' sport. Whenever a young woman or a handsome youth fell into their hands, they were torn to pieces by the violent struggles of those who tried to secure them, and this in the end drove the despoilers to kill one another. (*Hist.* 3.33)

Even granting considerable exaggeration within Tacitus' narrative, there was and today remains a tendency for conquering armies to assert political domination through sexual domination (one thinks of Guantanamo and the Mahmudiyah rape and killings). Since legionaries were prohibited from marriage, any sexual outlet was both surreptitious and illegal. Despite the laws, it would be absurd to suppose that soldiers remained celibate. Sex workers were commonly found near military camps, but during war there was little consequence for sexually assaulting conquered women and youth in the heat or immediate aftermath of battle. Such women had little hope of justice, as Roman administrators were unlikely to be sympathetic to their plight and, lacking Roman citizenship, they had few other means of recourse.

This rape culture was built into Greco-Roman sexual discourse, with binaries of dominant-submissive, active-passive, penetrating-penetrated, Roman-peregrine mapping almost perfectly onto one another and informing all sexual activity. It was hardly a stretch to include conqueror-conquered among these relations. David Halperin thus determines that "what was fundamental to [the Greco-Roman] experience of sex . . . was not anything we would regard as essentially sexual; it was rather the modality of power-relations that informed and structured the act."[19] The kings of the east thus humiliate the woman renowned among for her luxury and sexual promiscuity (18:3, 9). Revelation frames the degradation of the woman as an ironic reversal—the once-proud and wealthy woman has received her comeuppance—but the sexism implicit within this fantasy is disturbing. Some commentators object that this vision does not endorse violence against women as such, it merely uses the image of the woman as a means of criticizing Rome.[20] Jennifer Glancy and Stephen Moore caution against such a generous interpretation of Revelation, observing that sex workers did and continue to experience brutal violence at the hands of their clients; to treat her as a mere metaphor is to ignore the normalization of both physical and sexual violence against women.[21] John of Patmos never criticizes the vivid rendition of the violence meted upon her body, instead expecting the reader to relish the gory details.

Second, economic fallout from the destruction of Babylon/Rome is significant. The woman is depicted as seducing the kings of the world with her luxury. The economic import of Rome to the Mediterranean world is difficult to overstate, being both the wealthiest city in the empire, acquired largely through sea trade. The description by the second-century Roman orator Aristides, while romanticized (if giving a nod to the exploitation required for such economic centrality), is only a modest exaggeration:

> So many merchant ships arrive here, conveying every kind of goods from every people every hour and every day, so that the city is like a factory common to the whole earth. It is possible to see so many cargoes from India and even Arabia, Felix, if you wish, that one imagines that for the future the trees are left bare for the people there and that they must come here to beg for their own produce if they need anything. . . . The arrival of ships never stop, even for the sea. . . . So everything comes together here: trade, seafaring, farming, the scourings of the mines, all the crafts that exist or have ever existed. (*Orat.* 26)

What would be the significance of such a city's end? As a point of comparison, chapter 2 argued that the destruction of Jerusalem in the Jewish War led to massive shifts in the Palestinian economy: the loss of the temple cult and arrival of a much larger military garrison required many to adapt their labor to suit the new economy. Consequently, when the city of Rome is destroyed, Revelation imagines that the impact is felt most significantly by merchants (18:11–20). It is little surprise that John determines that "your merchants were the world's important people" (18:23). With the degree of devastation that Revelation imagines, the widespread death, looting, and destruction would surely diminish the capital city's economic standing.

Finally, implicit within all of this is the importance of the city's military destruction. Though it is never depicted, the ten kings and the beast bring about its devastation (17:16–17). These characters perform the will of God without knowing such. This is akin to the role of the Roman army's destruction of Jerusalem in Mark 13:1–37 (see chapter 3). Responsibility is attributed solely to the kings—it is unclear whether one should infer that they act on their own with superhuman abilities, or if they have armies at their command.

SEVEN KINGS AND THE CLIMACTIC BATTLE— REV 17:9–14, 19:11–21

One of the passages most transparently related to the Roman Empire in Revelation is its discussion of seven kings, culminating in the beast, and the ten kings who serve the beast, waging war against the Lamb. It is widely assumed that these kings correspond to specific emperors, but identifications is complicated and a widespread matter of disagreement: does Julius Caesar (under whom the Republic effectively ended) count as the first emperor, or is it Augustus (under whom the empire officially began)? What about the Year of the Four Emperors (69 CE), under whom there was a quick succession of insignificant emperors? Efforts to ascertain which sequence of kings have yet to yield a consensus. The matter is further complicated by Revelation's figurative approach to numbers—numbers such as seven and ten are laden with symbolic import; John may have contrived these numbers (cf. the artificial fourteen generations in Jesus' genealogy in Matt 1:1–17) or did not intend them to be taken literally at all.

Regardless, the battle itself draws heavily from existing biblical imagery, evoking Psalms' heavenly warrior (2:1–11), the exemplary ruler of Isa 63:1–6 (cf. Joel 3:11–13), and the wedding of a victorious king (Ps 45:3–4). But beyond biblical references, Revelation draws upon the image of the Roman triumph/procession, discussed above in connection with Jesus' entrance to Jerusalem and Paul's description of the Lord's arrival (cf. Figure 4.1). The comparison by R. Alastair Campbell is concise but thorough:

> In the same way Revelation depicts a procession led by a figure on a white horse, the symbol of victory, followed by his victorious, white-clad, soldiers, [is] also riding white horses. The fact that he is mounted and not in a chariot serves to bring out his oneness with those who, like him, have borne faithful witness. All are riding white horses. The rider is already crowned, as a *triumphator* should be. Like his Roman counterpart John's rider is dressed in a purple robe, but with the significant difference that the rider's robe owes its colour to its having been dipped in blood. Opinions differ as to whether this is the blood of his enemies or his own blood, but either way we should not miss the obvious point that if the rider's robe is dipped in blood it is because the battle is already over. His robe, suitably decorated, further declares him to be King of Kings and Lord of Lords.[22]

Some commentators are not entirely convinced by the parallels between the two. Objections stem from purported differences between Revelation and such triumphs: rather than riding a chariot, Christ rides a horse; he wears a diadem rather than a wreath; his robe is not purple but blood red, etc. These objections are inadequate for two reasons. First, there was not a single, uniform way in which triumphs proceeded. Rather, their performance varied from emperor to emperor, attending to their distinct interests and the myths they were trying to cultivate about themselves. Second, this objection operates on the assumption that denizens of the Roman Empire were well informed as to the proceedings of a triumph. As discussed above in tandem with Paul's eschatology, it is unlikely that a civilian would have witnessed a triumph with their own eyes. Rather, ancient writers were dependent upon stereotypical imagery and oral traditions, neither of which were especially reliable for recounting such images. Consequently, disparity between Revelation's description and the actual proceedings of Roman triumphs are attributable to a sort of "rounding error" that often happens when an author describes events that they little first-hand experience.

The heavenly army's garb is worthy of comment (19:13–14). The bloody robe of Christ is juxtaposed with the pure linen of his soldiers. It was common for soldiers to wear linen in hot or desert regions, such as Judaea and Egypt (e.g., §22, §167, §168), but was in use long before the Roman era; Greek and Etruscan soldiers used hardened linen as armor, attested as early as the Homeric epics (*Il.* 2.529, 2.830).[23] However, the linen of the heavenly armies does not seem to be an armor, but an otherworldly variety of linen clothing, connoted by their pure white quality. The battle itself between Christ and the evil forces is not so much a battle as a slaughter. After a description of the preparations for war—the beast and kings command a substantial army (19:19)—we have no impression that anything other than a one-sided bloodbath ensued.

CONCLUSION

The Book of Revelation's hostility to Rome is as thinly veiled as it could possibly be, devoting extensive discussion to Rome's destruction and humiliation. This is widely recognized among New Testament scholars. Typical of this understanding is Werner Kelber, who writes,

> In sum, the rogue city, the seductive Woman, and the abominable Beast are symbols with floating and interactive connotations whose core identity, however, is Rome, the paradigmatic city of imperial corruption and idolatry. Once we recognize that the Apocalypse's objection to Rome found expression through the principal trinity of evil representations, not by reference simply to the ancient city of Babylon, the profundity of its anti-Roman sentiments is difficult to overstate.[24]

Revelation's central target is Rome, the premier political power and oppressor of the Mediterranean. John articulates, obliquely but precisely, a utopian vision of life without Rome after a cataclysmic eschatological battle—life free of Roman cruelty and exploitation.

But while many see this as a subversive message of hope, it is difficult to avoid the impression that John of Patmos simply wishes to replace the Roman Empire with another empire. This "hope" is mitigated by the troubling misogyny and bloodthirsty fantasies, suggesting a more ambivalent stance to Roman violence. As Kimberly Stratton argues, John consistently reinscribes Roman imperial violence as part of its purportedly subversive message. "[Rev 19:11–16] presents Christ in the guise of heavenly emperor: he leads a vast cavalry that vanquishes the nations of the world; he has Imperium to judge and to make war; he wears the purple robe of royalty, dyed in the blood of his own martyrdom or of his enemies whom he has slaughtered, and bears the name King of kings—supplanting the Roman emperor as ruler of the world."[25] Revelation is thus not merely resisting Rome, but actively advancing other interests, interests that—while indeed anti-Roman—are inextricable of the horrific violence it criticizes. This cannot be resolved by pointing to a power imbalance (i.e., powerful Romans vs. oppressed Christians), as sexism is inherent to the vision it proffers. Thus, while Revelation may in some sense be a "critique from below," the force of its critique should not be separated from its own misogynist politics.

NOTES

1. Leif E. Vaage, *Borderline Exegesis*, Signifying (on) Scriptures (University Park: Pennsylvania State University Press, 2014), 126.

2. See the overview of scholarship in David E. Aune, *Revelation*, Word Biblical Commentary 52, 3 vols. (Dallas: Nelson, 1997–1998), 1.lvi–lxx; G. K. Beale, *The Book of Revelation*, New International Greek Testament Commentary (Louisville: Eerdmans, 1999), 4–27.

3. Christopher Webber, *The Thracians 700 BC–AD 46*, Men-at-Arms 360 (London: Osprey, 2001), 33–34. The identification of the *rhomphaia* is still disputed among military historians; for an overview of the controversy, see Duncan Head, "The Rhomphaia Lives!" *Slingshot* 77 (1978): 4–11. Most other proposals for the *rhomphaia* involve long, heavy, and single-edged weapons.

4. Robert H. Mounce, *The Book of Revelation*, New International Commentary on the New Testmaent, Revised ed. (Grand Rapids: Eerdmans, 1998), 60; Grant R. Osborne, *Revelation*, Baker Exegetical Commentary on the New Testament (Grand Rapids: Baker Academic, 2002), 140.

5. Craig R. Koester, *Revelation: A New Translation with Introduction and Commentary*, Anchor Bible Commentary 38A (New Haven: Yale University Press, 2014), 394–395, citing Plutarch *Crass.* 24.5–25.5, *Ant.* 34.3–5; Tacitus *Ann.* 15.1–19 on the Parthians and Ovid *Trist.* 5.7.13–15 on the Sarmatians and Getans.

6. These military units used the word *sagittariorum* in their unit name. *alae* and *cohortes equitatae* also included mounted archers.

7. Beale, *Revelation*, 399.

8. Bruce J. Malina and John J. Pilch, *Social-Science Commentary on the Book of Revelation* (Minneapolis: Fortress, 2000), 131–132.

9. See the overview of the issues at play in Beale, *Revelation*, 509.

10. The chimera is attested in, e.g., Homer *Il.* 6.181–182; Hesiod *Theog.* 319–324; Ovid *Trist.* 4.7.13; Seneca *Ep.* 113.9. This creature had a snake tail, a lion's head, and was known for breathing fire. See, e.g., Koester, *Revelation*, 467; Aune, *Revelation*, 2.539. While Revelation certainly evokes individual features of the chimera, the creatures in question are clearly described as horses, with features "like" these other animals—these body parts are not simply transposed onto horses.

11. Martin, *Hellenistic Religions*.

12. See, e.g., Steven J. Friesen, *Imperial Cults and the Apocalypse of John: Reading Revelation in Ruins* (Oxford: Oxford University Press, 2001), 43–55.

13. See, e.g., Brent D. Shaw, "The Myth of the Neronian Persecution," *JRS* 105 (2015): 73–100; Brent D. Shaw, "Response to Christopher Jones: The Historicity of the Neronian Persecution," *New Testament Studies* 64 (2018): 231–242; Heidi Wendt, "*Ea Superstitione*: Christian Martyrdom and the Religion of Freelance Experts," *JRS* 105 (2015): 183–202.

14. Adela Yarbro Collins, *Crisis and Catharsis: The Power of the Apocalypse* (Philadelphia: Westminster, 1984), 84. Emphasis in original.

15. Leonard L. Thompson, *The Book of Revelation: Apocalypse and Empire* (Oxford: Oxford University Press, 1990), 175.

16. Marius Heemstra, *The Fiscus Judaicus and the Parting of the Ways*, WUNT II 241 (Tübingen: Mohr Siebeck, 2010), 103–133.

17. Aune, *Revelation*, 3.919–928.

18. Koester, *Revelation*, 679; Richard Bauckham, *Climax of Prophecy: Studies on the Book of Revelation* (London: T&T Clark, 2000), 438.

19. David M. Halperin, "One Hundred Years of Homosexuality," *Diacritics* 16/2 (1986): 40.

20. E.g., Barbara Rossing, *The Choice Between Two Cities: Whore, Bride, and Empire in the Apocalypse*, Harvard Theological Studies 48 (Harrisburg: Trinity Press International, 1999), 87–90.

21. Jennifer A. Glancy and Stephen D. Moore, "How Typical a Roman Prostitute Is Revelation's 'Great Whore'?" *JBL* 130 (2011): 551–569.

22. R. Alastair Campbell, "Triumph and Delay: The Interpretation of Revelation 19:11–20:10," *Evangelical Quarterly* 80 (2008): 7.

23. See the extensive study in Gregory S. Aldrete, Scott M. Bartell, and Alicia Aldrete, *Reconstructing Ancient Linen Body Armor: Unraveling the Linothorax Mystery* (Baltimore: Johns Hopkins University Press, 2013).

24. Werner H. Kelber, "Roman Imperialism and Early Christian Scribality," in *Orality, Literacy and Colonialism in Antiquity*, ed. Jonathan A. Draper, SemeiaSt 47 (Atlanta: SBL, 2004), 151.

25. Kimberly B. Stratton, "The Eschatological Arena: Reinscribing Roman Violence in Fantasies of the End Times," *Biblical Interpretation* 17 (2009): 63.

Conclusion

Reading a Complicated Bible in Complicated Times

The New Testament lacks a single, consistent depiction of the military. On the one hand, Revelation criticizes Roman imperialism as directly as it can, fantasizing about its destruction in gory terms. Conversely, Luke-Acts minimizes tensions between Roman soldiers and early Christians, depicting positive and respectful interactions whenever possible. More complex examples are also found in texts that seem to have little interest in generalizing about "the military" as a category: Mark and Matthew deploy stereotypes of brutish lower-class soldiers and are more sympathetic to centurions as a potential follower of Christ. This is not to mention the many texts that have little interest in the military at all: 1–3 John, 1–2 Peter, James, Jude, etc. The New Testament not only offers both "yes" and "no" responses to the Roman military, but "who cares?" as well. We have seen that the depiction of the military vacillates depending on author and context, rendering the corpus we term the "New Testament" both ambiguous and ambivalent toward the military.

The New Testament's ambivalence to agents of state violence may be instructive for thinking about the politics of resistance today. Though some biblical writers clearly did experience the military as something analogous to the modern "war machine," other writers took it as part of daily life. In so doing, the military becomes part of the mundane, leading us to ask how Roman state violence, "in all its ordinariness," worked.[1] Beyond the physical violence that was common practice, the military also acted as a hidden generative force in cultural production; for instance, monetization and military-constructed infrastructure had both positive and negative effects for denizens of the Roman East. This goes doubly so for the Gospels, as the Jewish War and resulting trauma still linger throughout their stories—the loss of the temple, the influx of foreign soldiers, and an emphatically *Gentile* landscape are detectable concerns among the Synoptic Gospels. The military's violence was experienced in various ways, sometimes warranting condemnation, while others preferred to live with it as peacefully as possible. It is therefore important to push back against depictions of early Christian and Jewish discourse that imagine a dualistic world comprised of the empire and its collaborators on the one hand against a small but clearly-defined resistance on the other. Though there is an obvious appeal in this notion of "pure origins" (often locating its decline in the rise of Constantinian empire-building), it oversimplifies the inherent ambivalence of the military among early Christians.[2] Whatever posture the evangelists may have held toward the military, it forms an essential part of the Gospels both in terms of the events that prompted evangelists to write their texts, as well as the content of their stories.

How might one address biblical ambivalence towards state violence? One approach is to follow the route of Hector Avalos and use this data as an indictment of early Christians as self-serving collaborators with the powers that be.[3] By this standard, the absence of a cohesive and unambiguous anti-imperial message renders the New Testament a political failure: the texts of the New Testament create ample space for Roman violence and when they criticize such violence they only do so indirectly. Such reasoning seeks to dispute the unique origins of Christianity and its theological uniqueness: if Jesus' message is not pure, then current iterations of Christianity are even less so. This approach, however, is deeply problematic. This is not intended as an endorsement of settler colonialism, imperial formations, or state violence. Rather, this approach operates within a dualistic ethical framework that is unhelpful. To rephrase, I am not trying to argue that the New Testament promotes a "bad Jesus," but that it is the product of mundane human social interests.

We might thus see biblical ambivalence and complicity as reflective of our own situation. The frequent preoccupation with the "purity politics" of early Christianity obscures the ways in which daily life *required* complicity and how that complicity was variously navigated. The propensity to ignore or explain away such complicity, Slavoj Žižek suggests, is endemic to our present economic situation of neoliberal capitalism. Žižek argues that the smooth functioning of a given economic system is inherently violent, what he terms "objective-structural violence."[4] In short, he suggests treating such activity as though it were merely part of the "private/personal sphere" permits one to disavow their role in the exploitation of others' labor. Thus, whatever harm may come from one's economic actions, one is not encouraged to understand their actions as "violent" or as part of a broader violence. The way in which economic conditions are not understood as violent serves the capitalist liberal imaginary well:

> Liberalism's exclusion of economic relations from politics proper is an ideological buttress of capitalism. Where people are most affected by power—in relation to work, wages, and the material necessities of life—they are granted no decision-making faculty. We do not vote over our wages, work hours, unemployment status, benefits, or the accessibility of child care and health care—all in realms designated under capitalism as private.[5]

Or, for that matter, Ralph Waldo Emerson's indictment of bourgeois acceptance of exploitative consumption: "One plucks, one distributes, one eats. . . . Yet none feels himself accountable. He did not create the abuse; he cannot alter it; what is he? An obscure private person who must get his bread."[6] Thus, even though literally every computer uses parts produced by a company known for work conditions so dismal that it led to a suicide epidemic, the consumer rarely understands themselves to be morally implicated in such exploitation. It is precisely by keeping these issues at arm's length that exploitative economic systems are able to function smoothly; economic practices that are too easily perceived as "unjust" are liable to be interrupted, whether by work stoppage (see, e.g., *O.Ber.* 2.126), boycott (e.g., the Jewish refusal to purchase Roman pottery after a mass crucifixion in 4 BCE), or even revolt. The point Žižek makes with objective-structural violence is that even when an economic system appears to function smoothly, it is built upon invisible violence.

We might take ambivalence not as a matter of resignation that leads to political paralysis: resigning ourselves to living a less-than-perfect life, resigning to the fact that our ideals do not quite match our aspirations, etc. Rather, we may take it as an opportunity for further reflection and analysis of our role in these invisible forms of violence. The value of ambivalence and complicity is that they do not require us to be "outside" of violence to be in a place to critically engage it. Philosopher Alexis Shotwell articulates the problem with precision:

> *If* we want a world with less suffering and more flourishing, it would be useful to perceive complexity and complicity as the constitutive situation of our lives, rather than as things we should avoid. . . . To say we live in compromised times is to say that although most people aim to *not* cause suffering, destruction, and death, simply by living, buying things, throwing things away, we implicate ourselves in terrible effects on ecosystems and beings both near and far away from us. We are inescapably entwined and entangled with others, even when we cannot track or directly perceive this entanglement.[7]

Shotwell observes that this myth of political and ideological purity serves the interests of society's most powerful. To take a potent instance of our context, the obvious contradictions of late capitalism—philanthropic billionaires whose fortunes are only possible via exploitation of countless others being but a single example—are increasingly framed as only apparent in nature; the solutions to the "big problems" of today are found within capitalism itself. For example, such billionaires enable the consumer's role in ending exploitation by offering alternative methods of consumption. The consumer's role in the exploitation of Latin American coffee bean farmers is mitigated by purchasing fair-trade beans at a somewhat higher price from "ethical" coffee shops.[8] The consumer benefits by having a clean conscience, the farmer by working for fair wages, and the company by the selling to a class of consumers it may not have otherwise. Žižek, however, is unconvinced that such business models significantly ease exploitative processes. Indeed, Žižek contends that there is a collusive element to all of this, as the basic relations of exploitation remains uncontested: globalization, cash-crop farming, antiunion policies, the production of an amenable "ethical consumer" class, not to mention the continuation of the late capitalist means of production are all central to this model. Though philanthropic capitalists and ethical consumers claim to rectify the problems of capitalist injustice, they merely put a friendly face on inequitable modes of production; those producing the crops necessary for these goods receive only a fraction of their value as compensation, while excess value is extracted and remains in the hands of capitalists residing in the North Atlantic. If anything, late capitalism functions all the smoother because it ideologically subsumes previously dissenting classes into its own logic. Žižek offers a normative political assertion: sure, the farmers might live a life that is a little bit better because a given coffee company uses fair trade beans, but they ultimately remain in the same economic system that produces exploitative conditions. The myth of the "ethically pure" consumer, in this case, serves the interests of global capital, rendering invisible the violence required to sustain everyday life in the global north.

The question of ambivalence and complicity with imperial violence in New Testament texts may help us clarify these issues as well. Luke-Acts, for instance, has a clear chauvinism that favors the elite; in most cases, the narrator focuses upon the most privileged person in a scene at the expense of most others. Thus, Philip's interaction with the Ethiopian Eunuch requires the reader to forget that someone was needed to drive the chariot that the Eunuch and Philip were riding in.[9] Or, as mentioned above, the use of elitist stereotypes in Matthew and Mark when depicting the military, reducing low-ranking soldiers to bullies and raising some elites to the level of respectable and respectful men. Feminist scholars have done ample work on the issue of "texts of terror"—biblical texts that legitimate gender, ethnic, and other forms of violence—a category that does not seem quite appropriate in this instance. The narratives we are discussing do not endorse such violence per se, but participate in the common-but-pernicious act of "forgetting" the disparity of power that defines the relationship of the characters in these stories. This forgetting, I would contend, is a form of objective-structural violence, subtle as it may be.

The texts comprising the New Testament have been afforded an authoritative and normative status within much of the world, a status that has often led people to claim its message as their own—in different and competing ways. Jesus is simultaneously taken to legitimate pacifism, nonviolence, armed resistance, just war theories, and genocidal policies by those seeking to lend credence to their political projects. If there is no single, cohesive theory of violence in the New Testament, then perhaps it is time to abandon the quest for "biblical ethics" and explore the impurities of its texts so as to elucidate the problems of our own culture. What can we learn from our discomfort with Luke's cozy relationship with Roman state violence? How might the misogyny of Revelation's vision help us understand gender problems in anti-imperial political movements today? What might the fact that these texts—including those that were adamantly anti-Roman—about a man executed by Roman state violence came to be constitutive of western imperialism tell us about the limits of political critique?[10] I do not claim to have answers for these questions, but in "complicated times," where participation in state violence is unavoidable, the troubling aspects of biblical texts can be powerful prompts for further reflection.

Lest this seem a nihilistic or hopeless conclusion, Shotwell notes the benefits of rejecting a "politics of purity" in a passage that summarizes the orientation of the present publication.

[The present book] is "against purity" rather than *for* any of the many things I am indeed for because precisely one of my imperatives is to be *against* without predicting all the things there are to be *for*. . . . [John Holloway writes:] "We live in an unjust society and we wish it were not so: the two parts of this sentence are inseparable and exist in constant tension with each other." To invoke the foundational "no" of being against purity means that when we talk about impurity, implication, and compromise we are also foregrounding the fact that we are not all equally implicated in and responsible for the reprehensible state of the world.[11]

Thus, rather than arguing over whether biblical texts are being "misinterpreted" when they are taken to authorize something we may not like (e.g., genocide, patriarchy, endorsement of slavery), it may be better to set aside the question of their ethical purity altogether and accept that there are many things we might reasonably deem unethical in every biblical text. The dismissal of purity positions interpreters so as to ascertain the differing interests of such impurity and their relationship to the powers-that-be: that there is, for instance, a substantial difference between Mark, in the wake of the Jewish War's trauma, uses of stereotypes about soldiers as opposed to Luke-Acts encouraging collaboration between Christians and agents of Roman violence; that there is a difference between Paul—a man subjected to a great deal of state violence—using incidental military metaphors while largely remaining critical of the empire and Ephesians idealizing conquest and military masculinity as part of its endorsement of patriarchy; that there is a substantial difference between taking up arms to revolt against the Roman state and taking up arms as a part of state violence against Jews and other Palestinians. This approach to ambivalence can be fruitfully extended to contemporary contexts as well; how we think about the differing roles of violence in maintaining and contesting systems of oppression. In the words of Bruce Lincoln, such a perspective might encourage us "to view as immoral any discourse or practice that systematically operates to benefit the already privileged members of society at the expense of others, and [. . .] reserve the same judgment for any society that tolerates or encourages such discourses and practices."[12]

NOTES

1. The phrase comes from Pierre Bourdieu, *On Television*, trans. Priscilla Parkhurst Ferguson (New York: New, 1998), 21.

2. See, e.g., §§342–362. See also the influential discussion of the "Constantinian shift" and its narrative of ethical decline in Christian politics in John Howard Yoder, "Is There Such a Thing as Being Ready for Another Millennium?" in *The Future of Theology: Essays in Honor of Jürgen Moltmann*, ed. Miroslav Volf, Carmen Krieg, and Thomas Kucharz (Grand Rapids: Eerdmans, 1996), 63–69.

3. Hector Avalos, *The Bad Jesus: The Ethics of New Testament Ethics*, Bible in the Modern World 68 (Sheffield: Sheffield Phoenix, 2015), 151–178. For more on this, see Zeichmann, "Liberal Hermeneutics."

4. Slavoj Žižek, *Violence: Six Sideways Reflections*, Big Ideas//Small Books (New York: Picador, 2008).

5. Dana L. Cloud, *Control and Consolation in American Culture and Politics: Rhetorics of Therapy*, Rhetoric and Society (Thousand Oaks: Sage, 1998), 164.

6. Ralph Waldo Emerson, *The Collected Works of Ralph Waldo Emerson. Vol. I: Nature, Addresses, and Lectures* (Cambridge: Belknap, 1991), 148.

7. Alexis Shotwell, *Against Purity: Living Ethically in Compromised Times* (Minneapolis: University of Minnesota Press, 2016), 8. Emphasis in original.

8. Slavoj Žižek, *First as Tragedy, Then as Farce* (London: Verso, 2009), 53–54.

9. Zeichmann, "Gender Minorities."

10. See, e.g., Leif E. Vaage, "Why Christianity Succeeded (in) the Roman Empire," in *Religious Rivalries in the Early Roman Empire and the Rise of Christianity*, ed. Leif E. Vaage, Studies in Christianity and Judaism 18 (Waterloo: Wilfred Laurier University Press, 2006), 253–278.

11. Shotwell, *Against Purity*, 18–19. Emphasis added, quoting John Holloway, *Change the World without Taking Power: The Meaning of Revolution Today*, Get Political 8, New ed. (London: Pluto, 2010), 6–7.

12. Bruce Lincoln, *Emerging from the Chrysalis: Studies in Rituals of Women's Initiation*, 2nd ed. (Chicago: University of Chicago Press, 1991), 112.

Bibliography

Aldrete, Gregory S., Scott M. Bartell, and Alicia Aldrete. 2013. *Reconstructing Ancient Linen Body Armor: Unraveling the Linothorax Mystery.* Baltimore: Johns Hopkins University Press.

Alföldy, Géza. 1988. *The Social History of Rome.* Translated by David Braund and Frank Pollock. Revised ed. London: Routledge.

Allison, Dale C., Jr. 2005. *Studies in Matthew: Interpretation Past and Present.* Grand Rapids: Baker.

Alston, Richard. 1995. *Soldier and Society in Roman Egypt: A Social History.* London: Routledge.

Alt, Albrecht. 1930. "Limes Palaestinae." *Palästinajahrbuch* 26: 43–82.

Álvarez Cineira, David. 2013. "The Centurion's Statement (Mark 15:39): A *restitutio memoriae*." Pages 146–161 in *Jesus—Gestalt und Gestaltungen: Rezeptionen des Galiläers in Wissenschaft, Kirche und Gesellschaft. Festschrift für Gerd Theißen zum 70. Geburtstag.* Edited by Petra von Gemünden, David G. Horrell, and Max Küchler. Novum Testamentum et Orbis Antiquus 100. Göttingen: Vandenhoeck & Ruprecht.

Ando, Clifford. 2007. "The Army and the Urban Elite: A Competition for Power." Pages 359–378 in *A Companion to the Roman Army.* Edited by Paul Erdkamp. Blackwell Companions to the Ancient World. Malden: Blackwell.

Applebaum, Shimon. 1977. "Judaea as a Roman Province: The Countryside as a Political and Economic Factor." *Aufstieg und Niedergang der römischen Welt* 8: 355–396. Part 2, *Principat.* Edited by Hildegard Temporini and Wolfgang Haase. Berlin: De Gruyter.

———. 1989. *Judaea in Hellenistic and Roman Times: Historical and Archaeological Essays.* Studies in Judaism in Late Antiquity 40. Leuven: Brill Academic.

Arnal, William E. 2001. *Jesus and the Village Scribes: Galilean Conflicts and the Setting of Q.* Minneapolis: Fortress.

———. 2008. "The Gospel of Mark as Reflection on Exile and Identity." Pages 57–67 in *Introducing Religion: Essays in Honor of Jonathan Z. Smith.* Edited by Willi Braun and Russell T. McCutcheon. London: Equinox.

Arnal, William E. and Russell T. McCutcheon. 2013. *The Sacred Is the Profane: The Political Nature of "Religion."* Oxford: Oxford University Press.

Aubert, Jean-Jacques. 1995. "Policing the Countryside: Soldiers and Civilians in Egyptian Villages in the 3rd and 4th Centuries A.D." Pages 257–265 in *La hiérarchie (Rangordnung) de l'armée romaine sous le Haut-Empire*. Edited by Yann Le Bohec. De l'archéologie à l'histoire. Paris: De Boccard.

Aune, David E. 1997–1998. *Revelation*. Word Bible Comentary 52. 3 vols. Dallas: Nelson.

Avalos, Hector. 2015. *The Bad Jesus: The Ethics of New Testament Ethics*. Bible in the Modern World 68. Sheffield: Sheffield Phoenix.

Baker, Patricia Anne. 2000. *Medical Care for the Roman Army on the Rhine, Danube and British Frontiers in the First, Second and Early Third Centuries AD*. Ph.D. Thesis, Newcastle University.

Ball, Warwick. 2000. *Rome in the East: The Transformation of an Empire*. London: Routledge.

Barclay, John M. G. 1996. *Jews in the Mediterranean Diaspora: From Alexander to Trajan (323 BCE–117 CE)*. Edinburgh: T&T Clark.

Bauckham, Richard. 2000. *Climax of Prophecy: Studies on the Book of Revelation*. London: T&T Clark.

Bazzana, Giovanni B. 2015. *Kingdom of Bureaucracy: The Political Theology of Village Scribes in the Sayings Gospel Q*. Bibliotheca Ephemeridum Theologicarum Lovaniensium 274. Leuven: Peeters.

Beale, G. K. 1999. *The Book of Revelation*. New International Greek Testament Commentary. Louisville: Eerdmans.

Bermejo-Rubio, Fernando. 2019. "Was Pontius Pilate a Single-Handed Prefect? Roman Intelligence Sources as a Missing Link in the Gospel's Story." *Klio* 101: forthcoming.

Bernett, Monika. 2007. "Roman Imperial Cult in the Galilee." Pages 337–356 in *Religion, Ethnicity, and Identity in Ancient Galilee: A Region in Transition*. Edited by Jürgen Zangenberg, Harold W. Attridge, and Dale B. Martin. Wissenschaftliche Untersuchungen zum Neuen Testament I 210. Tübingen: Mohr Siebeck.

Bingham, Sandra. 2013. *The Praetorian Guard: A History of Rome's Elite Special Forces*. London: Tauris.

Birley, Anthony R. 1980. *The People of Roman Britain*. Berkeley: University of California Press.

Birley, Eric. 1961. *Roman Britain and the Roman Army: Collected Papers*. 2nd ed. London: Kendal.

———. 1986. "Before Diplomas, and the Claudian Reform." Pages 249–257 in *Heer und Integrationspolitik: Die romischen Militärdiplome als historische Quelle*. Edited by Werner Eck and Hartmut Wolff. Passauer historische Forschungen 2. Köln: Böhlau.

———. 1988. "Promotions and Transfers in the Roman Army II: The Centurionate." Pages 206–220 in *The Roman Army: Papers 1929–1986*. Edited by Eric Birley. Mavors Roman Army Researches 4. Amsterdam: Gieben.

Blagg, T. F. C. and Anthony C. King, eds. 1984. *Military and Civilian in Roman Britain: Cultural Relationships in a Frontier Province*. BAR British Series 136. Oxford: BAR.

Bligh, Philip H. 1968. "A Note on Huios Theou in Mark 15:39." *Expository Times* 80: 51–53.

Bourdieu, Pierre. 1998. *On Television*. Translated by Priscilla Parkhurst Ferguson. New York: New.

Bowens, Lisa M. 2016. "Investigating the Apocalyptic Texture of Paul's Martial Imagery in 2 Corinthians 4–6." *Journal for the Study of the New Testament* 39: 3–15.

Braund, David. 1984. *Rome and the Friendly King: The Character of Client Kingship*. London: Croom Helm.

Breeze, David. 1974. "The Career Structure below the Centurionate." *Aufstieg und Niedergang der römischen Welt* 1: 435–451. Part 2, *Principat*. Edited by Hildegard Temporini. Berlin: De Gruyter.

Brent, Allen. 1999. *The Imperial Cult and the Development of Church Order: Concepts and Images of Authority in Paganism and Early Christianity before the Age of Cyprian*. Supplements to Vigiliae Christianae 45. Leiden: Brill.

Breytenbach, Cilliers. 1990. "Paul's Proclamation and God's 'thriambos': Notes on 2 Corinthians 2:14–16b." *Neotestamentica* 24: 257–271.

Brink, Laurena Ann. 2014. "Going the Extra Mile: Reading Matt 5:41 Literally and Metaphorically." Pages 111–128 in *The History of Religions School Today: Essays on the New Testament and Related Ancient Mediterranean Texts*. Edited by Thomas R. Blanton, IV, Robert Matthew Calhoun, and Clare K. Rothschild. Wissenschaftliche Untersuchungen zum Neuen Testament I 340. Tübingen: Mohr Siebeck.

———. 2014. *Soldiers in Luke-Acts: Engaging, Contradicting and Transcending the Stereotypes*. Wissenschaftliche Untersuchungen zum Neuen Testament II 362. Tübingen: Mohr Siebeck.

Brown, Raymond E. 1994. *The Death of the Messiah. From Gethsemane to the Grave: A Commentary on the Passion Narratives in the Four Gospels*. 2 vols. Anchor Bible Reference Library. New York: Doubleday.

Campbell, R. Alastair. 2008. "Triumph and Delay: The Interpretation of Revelation 19:11 20:10." *Evangelical Quarterly* 80: 3–12.

Caragounis, Chrys C. 1974. "ΟΨΩΝΙΟΝ: A Reconsideration of Its Meaning." *Novum Testamentum* 16: 35–57.

Carreras Monfort, César. 2002. "The Roman Military Supply during the Principate: Transportation and Staples." Pages 70–89 in *The Roman Army and the Economy*. Edited by Paul Erdkamp. Amsterdam: Gieben.

Carter, Warren. 2015. "Cross-Gendered Romans and Mark's Jesus: Legion Enters the Pigs (Mark 5:1–20)." *Journal of Biblical Literature* 133: 139–155.

Cassidy, Richard J. 2001. *Paul in Chains: Roman Imprisonment and the Letters of St. Paul*. New York: Crossroad.

Catchpole, David R. 1984. "The 'Triumphal' Entry." Pages 319–334 in *Jesus and the Politics of His Day*. Edited by Ernst Bammel and C. F. D. Moule. Cambridge: Cambridge University Press.

Chancey, Mark A. 2005. *Greco-Roman Culture and the Galilee of Jesus*. Society for New Testament Studies Monograph Series 134. Cambridge: Cambridge University Press.

Charlesworth, James H. and Mordechai Aviam. 2014. "Reconstructing First-Century Galilee: Reflections on Ten Major Problems." Pages 103–137 in *Jesus Research: New Methodologies and Perceptions*. Edited by James H. Charlesworth. Princeton-Prague Symposia Series on the Historical Jesus 2. Grand Rapids: Eerdmans.

Christian, Michelle. 2017. "Calculating Acts: Luke's Numeracy in Context." Pages 219–239 in *Luke on Jesus, Paul, and Christianity: What Did He Really Know?* Edited by John S. Kloppenborg and Joseph Verheyden. Biblical Tools and Studies 29. Leuven: Peeters.

Christiansen, Erik. 1984. "On Denarii and Other Coin-Terms in the Papyri." *Zeitschrift für Papyrologie und Epigraphik* 54: 271–299.

Cloud, Dana L. 1998. *Control and Consolation in American Culture and Politics: Rhetorics of Therapy*. Rhetoric and Society. Thousand Oaks: Sage.

Cohen, Getzel M. 1972. "The Hellenistic Military Colony: A Herodian Example." *Transactions and Proceedings of the American Philological Association* 103: 83–95.

Collins, Adela Yarbro. 1984. *Crisis and Catharsis: The Power of the Apocalypse.* Philadelphia: Westminster.

———. 2007. *Mark: A Commentary.* Hermeneia. Minneapolis: Fortress.

Cook, John Granger. 2010. *Roman Attitudes Toward the Christians: From Claudius to Hadrian.* Wissenschaftliche Untersuchungen zum Neuen Testament I 261. Tübingen: Mohr Siebeck.

———. 2014. *Crucifixion in the Mediterranean World.* Wissenschaftliche Untersuchungen zum Neuen Testament 327. Tübingen: Mohr Siebeck.

Cotton, Hannah M. 1996. "Courtyard(s) in Ein-gedi: *P.Yadin* 11, 19, and 20 of the Babatha Archive." *Zeitschrift für Papyrologie und Epigraphik* 112: 197–201.

———. 2001. "Ein Gedi Between the Two Revolts." *Scripta Classica Israelica* 20: 139–154.

———. 2006. "The Impact of the Roman Army in the Province of Judaea/Syria Palaestina." Pages 393–407 in *The Impact of the Roman Army (200 BC–AD 476): Economic, Social, Political, Religious, and Cultural Aspects.* Edited by Lukas De Blois and Elio Lo Cascio. Impact of Empire 6. Leuven: Brill.

Crawford, Michael. 1970. "Money and Exchange in the Roman World." *Journal of Roman Studies* 60: 40–48.

Crossan, John Dominic. 2007. *God and Empire: Jesus against Rome, Then and Now.* San Francisco: HarperOne.

Crossan, John Dominic and Jonathan L. Reed. 2004. *In Search of Paul: How Jesus's Apostle Opposed Rome's Empire with God's Kingdom.* San Francisco: HarperSanFrancisco.

Cumont, Franz. 1956. *Oriental Religions in Roman Paganism.* 2nd ed. New York: Dover.

Dar, Shimon. 1986. *Landscape and Pattern: An Archaeological Survey of Samaria 800 B.C.E.–636 C.E.* 2 vols. BAR International Series 308. Oxford: BAR.

Davies, Gwyn and Jodi Magness. 2013. "Was a Roman Cohort Stationed at Ein Gedi?" *Scripta Classica Israelica* 32: 195–199.

Davies, Jeffrey L. 1984. "Soldiers, Peasants and Markets in Wales and the Marches." Pages 129–142 in *Military and Civilian in Roman Britain: Cultural Relationships in a Frontier Province.* Edited by T. F. C. Blagg and Anthony C. King. BAR British Series 136. Oxford: BAR.

———. 1997. "Native Producers and Roman Consumers: The Mechanisms of Military Supply in Wales from Claudius to Theodosius." Pages 267–272 in *Roman Frontier Studies 1995.* Edited by W. Groenman-van Waateringe, B. L. Van Beek, W. J. H. Willems, and S. L. Wynia. Oxbow Monograph Series 91. Oxford: Oxbow.

Davies, Roy W. 1989. *Service in the Roman Army.* Edinburgh: Edinburgh University Press.

de la Bédoyère, Guy. 2017. *Praetorian: The Rise and Fall of Rome's Imperial Bodyguard.* New Haven: Yale University Press.

Devijver, H. 1986. "Equestrian Officers from the East." Pages 109–225 in *Defence of the Roman and Byzantine East.* Edited by David L. Kennedy and Philip Freeman. Vol. 1 of 2. BAR International Series 297. Oxford: BAR.

———. 1989. "Equestrian Officers in the East." Pages 77–111 in *The Eastern Frontier of the Roman Empire.* Edited by David H. French and Chris S. Lightfoot. Vol. 1 of 2. BAR International Series 553. Oxford: BAR.

———. 1992. "The Geographical Origins of Equestrian Officers." Pages 109–128 in *The Equestrian Officers of the Roman Imperial Army.* Edited by H. Devijver. Vol. 2 of 2. Marvors Roman Army Researches 9. Stuttgart: Steiner.

Di Segni, Leah. 2002. "The Water Supply of Roman and Byzantine Palestine in Literary and Epigraphical Sources." Pages 37–67 in *The Aqueducts of Israel*. Edited by David Amit, Joseph Patrich, and Yizhar Hirschfeld. Journal of Roman Archaeology: Supplementary Series 46. Portsmouth: JRA.

Di Segni, Leah and Shlomit Weksler-Bdolah. 2012. "Three Military Bread Stamps from the Western Wall Plaza Excavations." *Atiqot* 70: 21–31.

Dobson, Brian. 1974. "The Significance of the Centurion and "Primipilaris" in the Roman Army and Administration." *Aufstieg und Niedergang der römischen Welt* 1: 392–434. Part 2, *Principat*. Edited by Hildegard Temporini. Berlin: De Gruyter.

Donahue, John R. and Daniel J. Harrington. 2002. *The Gospel of Mark*. Sacra Pagina 2. Collegeville: Liturgical.

Downing, Francis Gerald. 2015. "Dale Martin's Swords for Jesus: Shaky Evidence?" *Journal for the Study of the New Testament* 37: 326–333.

Drinkwater, John. 1978. "The Rise and Fall of the Gallic Iulii: Aspects of the Development of the Aristocracy of the Three Gauls under the Early Roman Empire." *Latomus* 37: 817–850.

Droge, Arthur J. 2011. "Ghostlier Demarcations: The "Gospels" of Augustus and Mark." *Early Christianity* 2: 335–355.

Duff, Paul Brooks. 1992. "The March of the Divine Warrior and the Advent of the Greco-Roman King: Mark's Account of Jesus' Entry into Jerusalem." *Journal of Biblical Literature* 111: 55–71.

Dyson, Stephen. 1985. *The Creation of the Roman Frontier*. Princeton: Princeton University Press.

Eck, Werner. 1984. "Zum konsularen Status von Judaea im frühen 2. Jh." *Bulletin of the American Society of Papyrologists* 22: 55–67.

el-Khouri, Lamia. 2008. "The Roman Countryside in North-west Jordan (63 BC–AD 324)." *Levant* 40: 71–87.

Elliott, Neil. 2004. "The Apostle Paul's Self-Presentation as Anti-Imperial Performance." Pages 67–88 in *Paul and the Roman Imperial Order*. Edited by Richard A. Horsley. Harrisburg: Trinity Press International.

———. 2008. "The Apostle Paul and Empire." Pages 97–116 in *In the Shadow of the Empire: Reclaiming the Bible as a History of Faithful Resistance*. Edited by Richard A. Horsley. Louisville: Westminster John Knox.

Elton, Hugh. 1996. *Frontiers of the Roman Empire*. Bloomington: Indiana University Press.

Emerson, Ralph Waldo. 1991. *The Collected Works of Ralph Waldo Emerson. Vol. I: Nature, Addresses, and Lectures*. Cambridge: Belknap.

Engberg-Pedersen, Troels. 2003. "Review of Horsley, *Hearing the Whole Story*." *Journal of Theological Studies* 54: 230–245.

Eubank, Nathan. 2014. "Dying with Power: Mark 15,39 from Ancient to Modern Interpretation." *Biblica* 85: 247–268.

Ferrill, Arther. 1991. *Roman Imperial Grand Strategy*. Publications of the Association of Ancient Historians 3. Lanham: University Press of America.

Février, P. A. 1979. "L'armée romaine et la construction des aqueducs." *Dossiers de l'Archéologie* 38: 88–93.

Fewster, Gregory P. 2014. "'Can I Have Your Autograph?': On Thinking about Pauline Authorship and Pseudepigraphy." *Bulletin for the Study of Religion* 43/3: 30–39.

Fiensy, David A. 1991. *The Social History of Palestine in the Herodian Period: The Land Is Mine.* Studies in the Bible and Early Christianity 20. Lewiston: Mellen.

Fiensy, David A. and Ralph K. Hawkins, eds. 2013. *The Galilean Economy in the Time of Jesus.* Early Christianity and Its Literature 11. Atlanta: SBL.

Fink, Robert O. 1966. "*P.Mich.* VII 422 (Inv. 4703): Betrothal, Marriage, or Divorce?" Pages 9–17 in *Essays in Honor of C. Bradford Welles.* Edited by Alan E. Samuel. American Studies in Papyrology 1. New Haven: ASP.

Fischer, Moshe. 2011. "Rome and Judaea during the First Century CE: A Strange *modus vivendi.*" Pages 143–156 in *Fines imperii–imperium sine fine?* Edited by Günther Moosbaur and Rainer Wiegels. Osnabrücker Forschungen zu Altertum und Antike-Rezeption 14. Rahden: Leidorf.

Fischer, Moshe, Benjamin Isaac, and Israel Roll. 1996. *Roman Roads in Judaea II: The Jaffa-Jerusalem Roads.* BAR International Series 628. Oxford: Tempvs Reparatvm.

Fitzmyer, Joseph A. 1985. *The Gospel of Luke: A New Translation and Commentary.* Anchor Bible Commentary 28A. Garden City: Doubleday.

Flexsenhar, Michael A., III. 2018. *Christians in Caesar's Household: The Emperor's Slaves in the Makings of Early Christianity,* Inventing Christianity. University Park: Pennsylvania State University Press.

Forni, Giovanni. 1953. *Il reclutamento delle legioni da Augusto a Diocleziano.* Pubblicazioni della Facoltà di Filosofia e Lettere della Università di Pavia 5. Milan: Bocca.

———. 1974. "Estrazione etnica e sociale dei soldati delle legioni nei primi tre secoli dell' Impero." *Aufstieg und Niedergang der römischen Welt* 1: 339–391. Part 2, *Principat.* Edited by Hildegard Temporini. Berlin: De Gruyter.

Fortna, Robert T. 1988. *The Fourth Gospel and Its Predecessor: From Narrative Source to Present Gospel.* Philadelphia: Augsburg Fortress.

France, R. T. 2007. *The Gospel of Matthew.* New International Greek Testament Commentary. Grand Rapids: Eerdmans.

Fredriksen, Paula. 2015. "Arms and the Man: A Response to Dale Martin's 'Jesus in Jerusalem: Armed and Not Dangerous.'" *Journal for the Study of the New Testament* 37: 312–325.

———. 2015. "Review of N.T. Wright, *Paul and the Faithfulness of God.*" *Catholic Biblical Quarterly* 77: 387–391.

Freeman, Philip. 1998. "On the Annexation of Provinces to the Roman Empire." *Classics Ireland* 5: 30–47.

Friedheim, Emmanuel. 2007. "The Religious and Cultural World of Aelia Capitolina: A New Perspective." *Oriental Archive* 75: 125–152.

Friesen, Steven J. 2001. *Imperial Cults and the Apocalypse of John: Reading Revelation in Ruins.* Oxford: Oxford University Press.

———. 2004. "Poverty in Pauline Studies: Beyond the New Consensus." *Journal for the Study of the New Testament* 26: 323–361.

———. 2010. "Paul and Economics: The Jerusalem Collection as an Alternative to Patronage." Pages 27–54 in *Paul Unbound: Other Perspectives on the Apostle.* Edited by Mark Douglas Given. Peabody: Hendrickson.

Fuhrmann, Christopher J. 2012. *Policing the Roman Empire: Soldiers, Administration, and Public Order.* Oxford: Oxford University Press.

Fusco, Vittorio. 1991. "Problems of Structure in Luke's Eschatology Discourse (Luke 21:7–36)." Pages 72–92, 225–232 in *Luke and Acts.* Edited by Gerald O'Collins and Gilberto Marconi. New York: Paulist.

Garnsey, Peter D. and Greg Woolf. 1989. "Patronage of the Rural Poor in the Roman World." Pages 153–170 in *Patronage in Ancient Society*. Edited by Andrew Wallace-Hadrill. Leicester-Nottingham Studies in Ancient Society. London: Routledge.

Gelardini, Gabriella. 2016. *Christus Militans: Studien zur politisch-militärischen Semantik im Markusevangelium vor dem Hintergrund des ersten jüdisch-römischen Krieges*. Novum Testamentum Supplements 165. Leiden: Brill.

Gichon, Mordechai. 2002. "45 Years of Research on the *Limes Palaestinae*: The Findings and Their Assessment in the Light of the Criticisms Raised (C1st–C4th)." Pages 185–206 in *Limes XVIII: Proceedings of the XVIIIth International Congress of Roman Frontier Studies*. Edited by Philip Freeman, Julian Bennett, Zbigniew T. Fiema, and Birgitta Hoffmann. Vol. 1 of 2. BAR International Series 1084. Oxford: BAR.

Gilliver, Kate. 2007. "The Augustan Reform and the Structure of the Imperial Army." Pages 183–200 in *A Companion to the Roman Army*. Edited by Paul Erdkamp. Blackwell Companions to the Ancient World. Malden: Blackwell.

Glancy, Jennifer A. and Stephen D. Moore. 2011. "How Typical a Roman Prostitute Is Revelation's 'Great Whore'?" *Journal of Biblical Literature* 130: 551–569.

Goldsmith, Raymond W. 1984. "An Estimate to the Annual Structure of the National Product of the Early Roman Empire." *Review of Income and Wealth* 30: 263–288.

Gombis, Timothy G. 2010. *The Drama of Ephesians: Participating in the Triumph of God*. Downers Grove: InterVarsity.

González Salinero, Raúl. 2003. "El servicio militar de los judíos en el ejército romano." *Aquila Legionis* 4: 45–91.

Gracey, M. H. 1986. "The Armies of Judean Client Kings." Pages 311–323 in *Defence of the Roman and Byzantine East*. Edited by David L. Kennedy and Philip Freeman. Vol. 1 of 2. BAR International 297. Oxford: BAR.

Graf, D. F. 1994. "The Nabataean Army and the *Cohortes Ulpiae Petraeorum*." Pages 265–305 in *The Roman and Byzantine Army in the East*. Edited by Edward Dąbrowa. Krakow: Drukarnia Uniwersytetu Jagiellońskiego.

Green, Joel B. 1997. *The Gospel of Luke*. New International Commentary on the New Testament. Grand Rapids: Eerdmans.

Groenman-van Waateringe, W. 1997. "Classical Authors and the Diet of Roman Soldiers: True or False?" Pages 261–266 in *Roman Frontier Studies 1995*. Edited by W. Groenman-van Waateringe, B. L. Van Beek, W. J. H. Willems, and S. L. Wyrria. Oxbow Monograph Series 91. Oxford: Oxbow.

Groh, Dennis E. 1998. "The Stratigraphic Chronology of the Galilean Synagogue from the Early Roman Period Through the Early Byzantine Period (ca. 420 C.E.)." Pages 51–69 in *Ancient Synagogues: Historical Analysis and Archaeological Discovery*. Edited by Dan Urman and Paul V. M. Flesher. Vol. 1 of 2. Studia Post-Biblica 47. Leiden: Brill.

Guthrie, George H. 2015. *2 Corinthians*. Baker Exegetical Commentary on the New Testament. Grand Rapids: Baker Academic.

Haensch, Rudolf. 2001. "Inschriften und Bevölkerungsgeschichte Niedergermaniens: Zu den Soldaten der *legiones I Minervia* und *XXX Ulpia Victrix*." *Kölner Jahrbuch* 33: 89–134.

Halperin, David M. 1986. "One Hundred Years of Homosexuality." *Diacritics* 16/2: 34–45.

Hamel, Gildas. 1990. *Poverty and Charity in Roman Palestine, First Three Centuries C.E.* Near Eastern Studies 23. Berkeley: University of California Press.

———. 2010. "Poverty and Charity." Pages 308–324 in *The Oxford Handbook of Jewish Daily Life in Roman Palestine*. Edited by Catherine Hezser. Oxford: Oxford University Press.

Hanson, A. E. 1989. "Village Officials at Philadelphia: A Model of Romanization in the Julio-Claudian Period." Pages 429–440 in *Egitto e storia antica dall'ellenismo all'età araba: Bilancio di un confronto.* Edited by L. Crisculo and G. Geraci. Bologna: Cooperativa Libriria Uniersitaria Editrice Bologna.

———. 2001. "Sworn Declaration to Agents from the Centurion Cattius Catullus: P.Col. Inv. 90 (*P.Thomas* 5)." Pages 91–97 in *Essays and Texts in Honor of J. David Thomas.* Edited by Traianos Gagos and Roger S. Bagnall. American Studies in Papyrology 42. Oakville: ASP.

Har-Peled, Misgav. 2013. *The Dialogical Beast: The Identification of Rome with the Pig in Early Rabbinic Literature.* Ph.D. dissertation, Johns Hopkins University, Baltimore.

Haynes, Ian. 2013. *Blood of the Provinces: The Roman Auxilia and the Making of Provincial Society from Augustus to the Severans.* Oxford: Oxford University Press.

Head, Duncan. 1978. "The Rhomphaia Lives!" *Slingshot* 77: 4–11.

Heemstra, Marius. 2010. *The Fiscus Judaicus and the Parting of the Ways.* Wissenschaftliche Untersuchungen zum Neuen Testament II 241. Tübingen: Mohr Siebeck.

Heilig, Christopher. 2017. *Paul's Triumph: Reassessing 2 Corinthians 2:14 in Its Literary and Historical Context.* Biblical Tools and Studies 27. Leuven: Peeters.

Hekster, Olivier J. 2015. *Emperors and Ancestors: Roman Rulers and the Constraints of Tradition.* Oxford Studies in Ancient Culture and Representation. Oxford: Oxford University Press.

Hobbs, T. R. 2001. "Soldiers in the Gospels: A Neglected Agent." Pages 328–348 in *Social-Scientific Models for Interpreting the Bible: Essays by the Context Group in Honor of Bruce J. Malina.* Edited by John J. Pilch. Biblical Interpretation Series 53. Leiden: Brill.

Hoey, Allan S. 1939. "Official Policy towards Oriental Cults in the Roman Army." *Transactions of the American Philosophical Association* 70: 456–481.

Hölbl, Günther. 2005. *Altägypten im Römischen Reich: der römische Pharao und seine Tempel. III. Heiligtümer und religiöses Leben in den ägyptischen Wüsten und Oasen.* Mainz: Zabern.

Holder, Paul A. 1980. *Studies in the Auxilia of the Roman Army from Augustus to Trajan.* BAR International Series 70. Oxford: BAR.

Holloway, John. 2010. *Change the World without Taking Power: The Meaning of Revolution Today.* New ed. Get Political 8. London: Pluto.

Horsley, Richard A. 1989. *The Liberation of Christmas: The Infancy Narratives in Social Context.* London: Continuum.

———. 1995. *Galilee: History, Politics, People.* Valley Forge: Trinity Press International.

———. 2001. *Hearing the Whole Story: The Politics of Plot in Mark's Gospel.* Louisville: Westminster John Knox.

———. 2003. *Jesus and Empire: The Kingdom of God and the New World Disorder.* Minneapolis: Fortress.

———. 2005. "'By the Finger of God': Jesus and Imperial Violence." Pages 51–80 in *Violence in the New Testament.* Edited by Shelly Matthews and E. Leigh Gibson. New York: T&T Clark.

Howgego, Christopher. 1985. *Greek Imperial Countermarks: Studies in the Provincial Coinage of the Roman Empire.* Royal Numismatic Society Special Publication 17. London: Royal Numismatic Society.

Incigneri, Brian J. 2003. *The Gospel to the Romans: The Setting and Rhetoric of Mark's Gospel.* Biblical Interpretation Series 65. Leiden: Brill.

Isaac, Benjamin. 1981. "The Decapolis in Syria: A Neglected Inscription." *Zeitschrift für Papyrologie und Epigraphik* 44: 67–74.

————. 1986. "Reflections on the Roman Army in the East." Pages 383–395 in *Defence of the Roman and Byzantine East.* Edited by David L. Kennedy and Philip Freeman. Vol. 2 of 2. BAR International Series 297. Oxford: BAR.

————. 1990. "Roman Administration and Urbanization." Pages 151–159 in *Greece and Rome in Eretz Irael: Collected Essays.* Edited by Aryeh Kasher, Uriel Rappaport, and Gideon Fuks. Jerusalem: Israel Exploration Society.

————. 1991. "The Roman Army in Judaea: Police Duties and Taxation." Pages 458–461 in *Roman Frontier Studies 1989.* Edited by Valerie A. Maxfield and M. J. Dobson. Exeter: University of Exeter Press.

————. 1992. *The Limits of Empire: The Roman Army in the East.* Revised ed. Oxford: Clarendon.

————. 2010. "Infrastructure." Pages 145–164 in *The Oxford Handbook of Jewish Daily Life in Roman Palestine.* Edited by Catherine Hezser. Oxford: Oxford University Press.

Iverson, Kelly R. 2011. "A Centurion's "Confession": A Performance-Critical Analysis of Mark 15:39." *Journal of Biblical Literature* 130: 329–350.

James, Simon. 1999. "The Community of the Soldiers: A Major Identity and Centre of Power in the Roman Empire." Pages 14–25 in *Proceedings of the Eighth Theoretical Roman Archaeology Conference.* Edited by Patricia Baker, Colin Forcey, Sophia Jundi, and Robert Witcher. Oxford: Oxbow.

————. "Soldiers and Civilians: Identity and Interaction in Roman Britain." Pages 77–89 in *Britons and Romans: Advancing an Archaeological Agenda.* Edited by Simon James and Martin Millett. Council for British Archaeology Research Report 125. York: Council for British Archaeology.

Jennings, Theodore W., Jr. and Tat-Siong Benny Liew. 2004. "Mistaken Identities but Model Faith: Rereading the Centurion, the Chap, and the Christ in Matthew 8:5–13." *Journal of Biblical Literature* 123: 467–494.

Jensen, Morten Hørning. 2010. *Herod Antipas in Galilee: The Literary and Archaeological Sources on the Reign of Herod Antipas and Its Socio-Economic Impact on Galilee.* 2nd ed. Wissenschaftliche Untersuchungen zum Neuen Testament II 215. Tübingen: Mohr Siebeck.

Johnson, Brian. 2011. "Crurifragium: An Intersection of History, Archaeology, and Theology in the Gospel of John." Pages 86–100 in *My Father's World: Celebrating the Life of Reuben G. Bullard.* Edited by John D. Wineland, Mark Ziese, and James Riley Estep. Eugene: Wipf and Stock.

Johnson, Earl S. 1987. "Is Mark 15:39 the Key to Mark's Christology?" *Journal for the Study of the New Testament* 31: 3–22.

————. 2000. "Mark 15,39 and the So-Called Confession of the Roman Centurion." *Biblica* 81: 406–413.

Johnson, Steven R., ed. 2002. *Q 7:1–10: The Centurion's Faith in Jesus' Word.* Documenta Q. Leuven: Peeters.

Judge, E. A. 2008. *The First Christians in the Roman World: Augustan and New Testament Essays.* Wissenschaftliche Untersuchungen zum Neuen Testament I 229. Tübingen: Mohr Siebeck.

Juel, Donald. 1997. "The Strange Silence of the Bible." *Interpretation* 51: 5–19.

Kagan, Kimberly. 2006. "Redefining Roman Grand Strategy." *Journal of Military History* 70: 333–362.

Kalantzis, George. 2012. *Caesar and the Lamb: Early Christian Attitudes on War and Military Service.* Eugene: Cascade.

Kasher, Aryeh. 1990. *Jews and Hellenistic Cities in Eretz Israel: Relations of the Jews in Eretz-Israel with the Hellenistic Cities during the Second Temple Period (332 BCE–70 CE)*. Texte und Studien zum antiken Judentum 21. Tübingen: Mohr Siebeck.

Katsari, Constantina. 2008. "The Monetization of the Roman Frontier Provinces." Pages 242–267 in *The Monetary Systems of the Greeks and Romans*. Edited by W. V. Harris. Oxford: Oxford University Press.

Kelber, Werner H. 2004. "Roman Imperialism and Early Christian Scribality." Pages 135–153 in *Orality, Literacy and Colonialism in Antiquity*. Edited by Jonathan A. Draper. Semeia Studies 47. Atlanta: SBL.

Kennedy, David L. 2004. *The Roman Army in Jordan*. 2nd ed. London: Council for British Research in the Levant.

Kilpatrick, G. D. 1945. "Acts vii.52 ΕΛΕΥΣΙΣ." *Journal of Theological Studies* 46: 136–145.

———. 1958. "The Transmission of the New Testament and Its Reliability." *Bible Translator* 9: 127–136.

Kim, Seyoon. 2008. *Christ and Caesar: The Gospel and the Roman Empire in the Writings of Paul and Luke*. Grand Rapids: Eerdmans.

King, Anthony C. 1984. "Animal Bones and the Dietary Identity of Military and Civilian Groups in Roman Britain, Germany and Gaul." Pages 187–217 in *Military and Civilian in Roman Britain: Cultural Relationships in a Frontier Province*. Edited by T. F. C. Blagg and Anthony C. King. BAR British Series 136. Oxford: BAR.

Kittel, Gerhard. 1964. αἰχμάλωτος, αἰχμαλωτίζω, αἰχμαλωτεύω, αἰχμαλωσία, συναιχμάλωτός. In *Theological Dictionary of the New Testament*, edited by Gerhard Kittel. Grand Rapids: Eerdmans.

Klauck, Hans-Josef. 2006. *Ancient Letters and the New Testament: A Guide to Context and Exegesis*. Translated by Daniel P. Bailey. Waco: Baylor University Press.

Klinghardt, Matthias. 2007. "Legionsschweine in Gerasa: Lokalkolorit und historischer Hintergrund von Mk 5,1–20." *Zeitschrift für die Neutestamentliche Wissenschaft* 98: 28–48.

Kloppenborg, John S. 1992. "*Exitus clari viri*: The Death of Jesus in Luke." *Toronto Journal of Theology* 8: 106–120.

———. 2005. "*Evocatio Deorum* and the Date of Mark." *Journal of Biblical Literature* 124: 419–450.

———. 2011. "The Representation of Violence in Synoptic Parables." Pages 323–351 in *Mark and Matthew I: Comparative Settings: Understanding the Earliest Gospels in Their First Century Settings*. Edited by Eve-Marie Becker and Anders Runesson. Wissenschaftliche Untersuchungen zum Neuen Testament I 271. Tübingen: Mohr Siebeck.

Koester, Craig R. 2014. *Revelation: A New Translation with Introduction and Commentary*. Anchor Bible Commentary 38A. New Haven: Yale University Press.

Koester, Helmut. 1997. "Imperial Ideology and Paul's Eschatology in 1 Thessalonians." Pages 158–166 in *Paul and Empire: Religion and Power in Roman Imperial Society*. Edited by Richard A. Horsley. Harrisburg: Trinity Press International.

Koskenniemi, Erkki. 2009. *The Exposure of Infants among Jews and Christians in Antiquity*. Social World of Biblical Antiquity, Second Series 4. Sheffield: Sheffield Phoenix.

Krause, Jens-Uwe. 1996. *Gefängnisse im römischen Reich*. Heidelberger althistorische Beiträge und epigraphische Studien 23. Stuttgart: Steiner.

Kreissig, Heinz. 1970. *Die sozialen Zusammenhänge des judäischen Krieges: Klassen und Klassenkampf im Palästina des 1. Jahrhunderts v.u.Z*. Schriften zur Geschichte und Kultur der Antike 1. Berlin: Akademie.

Kyrychenko, Alexander. 2014. *The Roman Army and the Expansion of the Gospel: The Role of the Centurion in Luke-Acts.* Beihefte zur Zeitschrift für die neutestamentliche Wissenschaft 203. Berlin: De Gruyter.

Lambrecht, Jan. 1979. "Paul's Farewell-Address at Miletus: Acts 20,17–38." Pages 307–337 in *Les Actes des Apôtres: Traditions, Rédaction, Théologie.* Edited by Jacob Kremer. Bibliotheca Ephemeridum Theologicarum Lovaniensium 48. Leuven: Leuven University Press.

Lapin, Hayim. 2017. "Jerusalem the Consumer City: Temple, Cult, and Consumption in the Second Temple Period." Pages 241–254 in *Expressions of Cult in the Southern Levant in the Greco-Roman Period: Manifestations in Text and Material Culture.* Edited by Oren Tal and Zeev Weiss. Contextualizing the Sacred 6. Turnhout: Brepols.

Lau, Markus. 2007. "Die *Legio X Fretensis* und der Besessene von Gerasa: Anmerkungen zur Zahlenangabe „ungefähr Zweitausend" (Mk 5,13)." *Biblica* 88: 351–364.

Leander, Hans. 2013. *Discourses of Empire: The Gospel of Mark from a Postcolonial Perspective.* Semeia Studies 71. Atlanta: SBL.

Lewis, Naphtali. 1989. *The Documents from the Bar-Kochba Period in the Cave of Letters: Greek Papyri.* Judean Desert Studies 2. Jerusalem: Israel Exploration Society.

Lincoln, Bruce. 1991. *Emerging from the Chrysalis: Studies in Rituals of Women's Initiation.* 2nd ed. Chicago: University of Chicago Press.

Llewelyn, S. R. 1998. "The Career of T. Mucius Clemens and Its Jewish Connections." Pages 152–155 in *New Documents Illustrating Early Christianity 8.* Edited by S. R. Llewelyn. Grand Rapids: Eerdmans.

Long, Fredrick J. 2013. "*Ephesianoi* Paul's Political Theology in Greco-Roman Political Context." Pages 255–309 in *Christian Origins and Classical Culture: Social and Literary Contexts for the New Testament.* Edited by Stanley E. Porter and Andrew W. Pitts. Texts and Editions for New Testament Study 9. Leiden: Brill.

Luttwak, Edward N. 1976. *The Grand Strategy of the Roman Empire from the First Century A.D. to the Third.* 1st ed. Baltimore: Johns Hopkins University Press.

———. 1999. "Give War a Chance." *Foreign Affairs* 78/4: 36–44.

———. 2016. *The Grand Strategy of the Roman Empire from the First Century A.D. to the Third.* Revised and Updated ed. Baltimore: Johns Hopkins University Press.

MacDonald, Dennis R. 1983. *The Legend and the Apostle: The Battle for Paul in Story and Canon.* Louisville: Westminster John Knox.

Mack, Burton L. 1988. *A Myth of Innocence: Mark and Christian Origins.* Philadelphia: Fortress.

———. 1995. *Who Wrote the New Testament? The Making of the Christian Myth.* San Francisco: HarperSanFrancisco.

Macky, Peter W. 1998. *St. Paul's Cosmic War Myth: A Military Version of the Gospel.* Westminster College Library of Biblical Symbolism 2. Frankfurt: Lang.

MacMullen, Ramsay. 1959. "Roman Imperial Building in the Provinces." *Harvard Studies in Classical Philology* 64: 207–235.

———. 1966. *Soldier and Civilian in the Later Roman Empire.* Cambridge: Harvard University Press.

Magness, Jodi. 2002. "In the Footsteps of the Tenth Roman Legion in Judea." Pages 189–212 in *The First Jewish Revolt: Archaeology, History, and Ideology.* Edited by Andrea M. Berlin and J. Andrew Overman. London: Routledge.

Maier, Harry O. 2013. *Picturing Paul in Empire: Imperial Image, Text and Persuasion in Colossians, Ephesians and the Pastoral Epistles.* London: Bloomsbury.

Malherbe, Abraham J. 1983. "Antisthenes and Odysseus, and Paul at War." *Harvard Theological Review* 76: 143–173.

Malina, Bruce J. and John J. Pilch. 2000. *Social-Science Commentary on the Book of Revelation.* Minneapolis: Fortress.

Mann, J. C. 1979. "Power, Force and the Frontiers of the Empire." *Journal of Roman Studies* 69: 175–183.

———. 1983. *Legionary Recruitment and Veteran Settlement during the Principate.* Institute of Archaeology Occasional Publication 7. London: Institute of Archaeology.

———. 1996. *Britain and the Roman Empire: Collected Studies.* Brookfield: Variorum.

Martin, Dale B. 2014. "Jesus in Jerusalem: Armed and Not Dangerous." *Journal for the Study of the New Testament* 37: 3–24.

Martin, Luther H. 1987. *Hellenistic Religions: An Introduction.* Oxford: Oxford University Press.

Mason, Hugh J. 1974. *Greek Terms for Roman Institutions: A Lexicon and Analysis.* American Studies in Papyrology 13. Toronto: Hakkert.

Mason, Steve. 2008. *Judean War 2: Translation and Commentary.* Brill Josephus Project 1b. Leiden: Brill.

———. 2016. *A History of the Jewish War: AD 66–74.* Key Conflicts of Classical Antiquity. Cambridge: Cambridge University Press.

Mattern, Susan P. 1999. *Rome and the Enemy: Imperial Strategy in the Principate.* Berkeley: University of California Press.

Mattingly, David J. 2011. *Imperialism, Power, and Identity: Experiencing the Roman Empire.* Miriam S. Balmuth Lectures in Ancient History and Archaeology. Princeton: Princeton University Press.

Meggitt, Justin J. 1998. *Paul, Poverty and Survival.* Studies of the New Testament and Its World. Edinburgh: T&T Clark.

Mélèze-Modrzejewski, Joseph. 1995. *The Jews of Egypt: From Ramses II to Emperor Hadrian.* Translated by Robert Cornman. Philadelphia: Jewish Publication Society.

Metzger, Bruce M. 1994. *A Textual Commentary on the Greek New Testament.* 2nd ed. New York: United Bible Societies.

Michalak, Aleksander R. 2012. *Angels as Warriors in Late Second Temple Jewish Literature.* Wissenschaftliche Untersuchungen zum Neuen Testament II 330. Tübingen: Mohr Siebeck.

Millar, Fergus. 1982. "Emperors, Frontiers and Foreign Relations, 31 B.C. to A.D. 378." *Britannia* 13: 1–23.

———. 1993. *The Roman Near East: 37 BC–AD 337.* Carl Newell Jackson Lectures. Cambridge: Harvard University Press.

Miller, Amanda C. 2014. *Rumors of Resistance: Status Reversals and Hidden Transcripts in the Gospel of Luke.* Emerging Scholars. Minneapolis: Fortress.

Mitchell, Stephen. 1999. "The Cult of Theos Hypsistos between Pagans, Jews and Christians." Pages 81–146 in *Pagan Monotheism in Late Antiquity.* Edited by Polymnia Athanassiadi and Michael Frede. Oxford: Oxford University Press.

Mounce, Robert H. 1998. *The Book of Revelation.* Revised ed. New International Commentary on the New Testament. Grand Rapids: Eerdmans.

Nicolotti, Andrea. 2017. "The Scourge of Jesus and the Roman Scourge." *Journal for the Study of the Historical Jesus* 15: 1–59.

Noethlichs, Karl-Leo. 2000. "Der Jude Paulus: Ein Tarser und Römer?" Pages 53–84 in *Rom und das himmlische Jerusalem: Die frühen Christen zwischen Anpassung und Ablehnung.* Edited by Raban von Haehling. Darmstadt: Wissenschaftliche Buchgesellschaft.

Nolland, John. 2010. "'The Times of the Nations' and a Prophetic Pattern in Luke 21." Pages 133–147 in *Biblical Interpretation in Early Christian Gospels*. Edited by Thomas R. Hatina. Vol. 3. Library of New Testament Studies 376. London: T&T Clark.

Nossov, Konstantin. 2011. *Gladiator: The Complete Guide to Ancient Rome's Bloody Fighters*. Guilford: Lyons.

Osborne, Grant R. 2002. *Revelation*. Baker Exegetical Commentary on the New Testament. Grand Rapids: Baker Academic.

Oudshoorn, Jacobine G. 2007. *Roman and Local Law in the Babatha and Salome Komaise Archives: General Analysis and Three Case Studies on Law of Succession, Guardianship and Marriage*. Studies on the Desert of Judah 69. Leiden: Brill.

Patrich, Joseph and David Amit. 2002. "The Aqueducts of Israel: An Introduction." Pages 9–20 in *The Aqueducts of Israel*. Edited by David Amit, Joseph Patrich, and Yizhar Hirschfeld. Journal of Roman Archaeology: Supplementary Series 46. Portsmouth: JRA.

Pažout, Adam. 2015. "Spatial Analysis of Early Roman Fortifications in Northern Negev." Diploma thesis, Faculty of Arts, Charles University in Prague.

Peachin, Michael. 1999. "Five Vindolanda Tablets, Soldiers, and the Law." *Tyche* 14: 223–235.

———. 2007. "Petition to a Centurion from the NYU Papyrus Collection and the Question of Informal Adjudication Performed by Centurions (*P.Sijp* 15)." Pages 79–97 in *Papyri in Memory of P. J. Sijpesreijn (P.Sijp.)*. Edited by A. J. B. Sirks and K. Worp. American Studies in Papyrology 40. Oakville: ASP.

Perkins, Pheme. 1995. "The Gospel of Mark: Introduction, Commentary and Reflections." Pages 509–733 in *The New Interpreter's Bible*. Vol. 8 of 12. Nashville: Abingdon.

Pervo, Richard I. 1987. *Profit with Delight: The Literary Genre of the Acts of the Apostles*. Philadelphia: Fortress.

———. 2006. *Dating Acts: Between the Evangelists and the Apologists*. Sonoma: Polebridge.

———. 2009. *Acts: A Commentary*. Hermeneia. Minneapolis: Fortress.

Peters, Edward M. 1995. "Prison Before the Prison: The Ancient and Medieval Worlds." Pages 3–47 in *The Oxford History of the Prison: The Practice of Punishment in Western Society*. Edited by Norval Morris and David J. Rothman. Oxford: Oxford University Press.

Phang, Sara Elise. 2001. *The Marriage of Roman Soldiers, (13 B.C.–A.D. 235): Law and Family in the Imperial Army*. Columbia Studies in the Classical Tradition 24. Leiden: Brill.

———. 2008. *Roman Military Service: Ideologies of Discipline in the Late Republic and Early Principate*. Cambridge: Cambridge University Press.

Pollard, Nigel. 1996. "The Roman Army as a "Total Institution" in the Near East? Dura-Europos as a Case Study." Pages 211–228 in *The Roman Army in the East*. Edited by David L. Kennedy. Journal of Roman Archaeology Supplement 18. Ann Arbor: JRA.

———. 2000. *Soldiers, Cities, and Civilians in Roman Syria*. Ann Arbor: University of Michigan Press.

Price, S. R. F. 1984. "Gods and Emperors: The Greek Language of the Roman Imperial Cult." *Journal of Hellenic Studies* 104: 79–95.

Pucci Ben Zeev, Miriam. 2000. "L. Tettius Crescens' *Expeditio Iudaea*." *Zeitschrift für Papyrologie und Epigraphik* 133: 256–258.

Rapske, Brian. 1994. *The Book of Acts and Paul in Roman Custody*. The Book of Acts in Its First Century Setting 3. Grand Rapids: Eerdmans.

Reed, Jonathan L. 2000. *Archaeology and the Galilean Jesus: A Re-examination of the Evidence*. Harrisburg: Trinity Press International.

Richardson, Peter. 1996. *Herod: King of the Jews and Friend of the Romans*. Studies on Personalities of the New Testament. Edinburgh: T&T Clark.

———. 2004. *Building Jewish in the Roman East*. Supplements to the Journal for the Study of Judaism 92. Leiden: Brill.

Richmond, I. A. 1962. "The Roman Siege-Works of Masada, Israel." *Journal of Roman Studies* 52: 142–153.

Rocca, Samuel. 2008. *The Forts of Judaea 168 BC–73 AD: From the Maccabees to the Fall of Masada*. Fortress 65. Oxford: Osprey.

———. 2008. *Herod's Judaea: A Mediterranean State in the Classical World*. Texts and Studies in Ancient Judaism 122. Tübingen: Mohr Siebeck.

———. 2009. *The Army of Herod the Great*. Men-at-Arms 443. Oxford: Osprey.

———. 2010. "Josephus, Suetonius, and Tacitus on the Military Service of the Jews of Rome: Discrimination or Norm?" *Italia* 20: 7–30.

Rollens, Sarah E. 2015. "Review of Fiensy and Hawkins, eds., *The Galilean Economy in the Time of Jesus*." *Review of Biblical Literature* 2015/3: [n.p.].

Roller, Duane W. 1998. *The Building Program of Herod the Great*. Berkeley: University of California Press.

Rosenberger, Mayer. 1978. *The Coinage of Eastern Palestine and Legionary Countermarks, Bar-Kochba Overstrucks*. Jerusalem: Rosenberger.

Rosenthal-Heginbottom, Renate. 2008. "The Material Culture of the Roman Army." Pages 90–108 in *The Great Revolt in the Galilee*. Edited by Ofra Guri-Rimon. Haifa Museum 28. Haifa: University of Haifa.

Rossing, Barbara. 1999. *The Choice Between Two Cities: Whore, Bride, and Empire in the Apocalypse*. Harvard Theological Studies 48. Harrisburg: Trinity Press International.

Roth, Jonathan P. 1995. "The Length of the Siege of Masada." *Scripta Classica Israelica* 14: 87–110.

———. 1999. *The Logistics of the Roman Army at War (264 B.C.–A.D. 235)*. Columbia Studies in the Classical Tradition 23. Leiden: Brill.

———. 2002. "The Army and the Economy in Judaea and Palaestina." Pages 375–397 in *The Roman Army and the Economy*. Edited by Paul Erdkamp. Amsterdam: Gieben.

———. 2006. "Jews and the Roman Army: Perceptions and Realities." Pages 409–420 in *The Impact of the Roman Army (200 BC–AD 476): Economic, Social, Political, Religious, and Cultural Aspects*. Edited by Lukas de Blois and Elio Lo Cascio. Impact of Empire 6. Leuven: Brill.

———. forthcoming. "Jewish Military Forces in the Roman Service." Pages forthcoming in *Essential Essays for the Study of the Military in New Testament Palestine*. Edited by Christopher B. Zeichmann. Eugene: Pickwick.

Roxan, Margaret M. 1991. "Women on Frontiers." Pages 462–467 in *Roman Frontier Studies 1989*. Edited by Valerie A. Maxfield and M. J. Dobson. Exeter: University of Exeter Press.

Russell, James. 1995. "A Roman Military Diploma from Rough Cilicia." *Bonner Jahrbücher* 195: 67–133.

Saddington, Denis B. 1982. *The Development of the Roman Auxiliary Forces from Caesar to Vespasian (49 B.C.–A.D. 79)*. Harare: University of Zimbabwe Press.

———. 1996. "Roman Military and Administrative Personnel in the New Testament." *Aufstieg und Niedergang der römischen Welt* 26.3: 2409–2435. Part 2, *Principat*. Edited by Wolfgang Haase. Berlin: De Gruyter.

———. 2006. "The Centurion in Matthew 8:5–13: Consideration of the Proposal of Theodore W. Jennings, Jr., and Tat-Siong Benny Liew." *Journal of Biblical Literature* 125: 140–142.

Safrai, Ze'ev. 1992. "The Roman Army in the Galilee." Pages 103–114 in *The Galilee in Late Antiquity*. Edited by Lee I. Levine. New York: Jewish Theological Seminary of America.

———. 1994. *The Economy of Roman Palestine*. London: Routledge.

Sanders, E. P. 2002. "Jesus' Galilee." Pages 3–41 in *Fair Play: Diversity and Conflicts in Early Christianity. Essays in Honour of Heikki Räisänen*. Edited by Ismo Dunderberg, Christopher M. Tuckett, and Kari Syreeni. Novum Testamentum Supplements 103. Köln: Brill.

Sartre, Maurice. 2005. *The Middle East under Rome*. Translated by Catherine Porter and Elizabeth Rawlings. Cambridge: Harvard University Press.

Scheidel, Walter. 2006. "Stratification, Deprivation and Quality of Life." Pages 40–59 in *Poverty in the Roman World*. Edited by Margaret Atkins and Robin Osborne. Cambridge: Cambridge University Press.

Schellenberg, Ryan S. 2015. "The First Pauline Chronologist? Paul's Itinerary in the Letters and in Acts." *Journal of Biblical Literature* 134: 193–213.

Schoenfeld, Andrew J. 2006. "Sons of Israel in Caesar's Service: Jewish Soldiers in the Roman Military." *Shofar* 24/3: 115–126.

Schulten, Adolf. 1933. "Masada: Die Burg des Herodes und die Römischen Lager." *Zeitschrift des Deutschen Palästina-Vereins* 56: 1–179.

Schultz, Brian. 2007. "Jesus as Archelaus in the Parable of the Pounds (Lk. 19:11–27)." *Novum Testamentum* 49: 105–127.

Schürer, Emil. 1973–1987. *The History of the Jewish People in the Age of Jesus Christ (175 B.C.–A.D. 135)*. Translated by Geza Vermes, Fergus Millar, and Matthew Black. Revised English ed. 3 vols. Edinburgh: T&T Clark.

Schwartz, Daniel R. 2004. "Did the Jews Practice Infant Exposure and Infanticide in Antiquity?" *Studia Philonica Annual* 16: 61–95.

Schwartz, Saundra. 2003. "The Trial Scene in the Greek Novels and in Acts." Pages 105–137 in *Contextualizing Acts: Lukan Narrative and Greco-Roman Discourse*. Edited by Todd Penner and Caroline Vander Stichele. Society of Biblical Literature Symposium Series 20. Atlanta: SBL.

Schwartz, Seth. 1990. *Josephus and Judaean Politics*. Columbia Studies in the Classical Tradition 18. Leiden: Brill.

Segal, Arthur. 1994. *Theatres in Roman Palestine and Provincia Arabia*. Mnemosyne Supplementum 140. Leiden: Brill.

Shatzman, Israel. 1983. "The Beginning of the Roman Defensive System in Judaea." *American Journal of Ancient History* 8: 130–160.

———. 1991. *The Armies of the Hasmonaeans and Herod: From Hellenistic to Roman Frameworks*. Texte und Studien zum Antiken Judentum 25. Tübingen: Mohr Siebeck.

Shaw, Brent D. 1983. "Soldiers and Society: The Army in Numidia." *Opus* 2: 133–159.

———. 2015. "The Myth of the Neronian Persecution." *Journal of Roman Studies* 105: 73–100.

———. 2018. "Response to Christopher Jones: The Historicity of the Neronian Persecution." *New Testament Studies* 64: 231–242.

Shea, Christine R. 2013. "Names in Acts: A Cameo Essay." Pages 22–24 in *Acts and Christian Beginnings: The Acts Seminar Report*. Edited by Dennis E. Smith and Joseph B. Tyson. Salem: Polebridge.

Shean, John F. 2010. *Soldiering for God: Christianity and the Roman Army*. History of Warfare 61. Leiden: Brill.

Sherwin-White, Adrian N. 1963. *Roman Society and Roman Law in the New Testament*. Oxford: Oxford University Press.

———. 1984. *Roman Foreign Policy in the East*. Norman: University of Oklahoma Press.

Shiner, Whitney T. 2000. "The Ambiguous Pronouncement of the Centurion and the Shrouding of Meaning in Mark." *Journal for the Study of the New Testament* 22: 3–22.

Shotwell, Alexis. 2016. *Against Purity: Living Ethically in Compromised Times*. Minneapolis: University of Minnesota Press.

Sider, Ronald J. 2012. *The Early Church on Killing: A Comprehensive Sourcebook on War, Abortion, and Capital Punishment*. Grand Rapids: Baker Academic.

Smallwood, E. Mary. 1981. *The Jews under Roman Rule from Pompey to Diocletian: A Study in Political Relations*. 2nd ed. Studies in Judaism in Late Antiquity 20. Leiden: Brill.

Smith, Dennis E. and Joseph B. Tyson, eds. 2013. *Acts and Christian Beginnings: The Acts Seminar Report*. Salem: Polebridge.

Smith, Julien. 2011. *Christ the Ideal King*. Wissenschaftliche Untersuchungen zum Neuen Testament II 313. Tübingen: Mohr Siebeck.

Spaul, John. 2000. *Cohors²: The Evidence for and a Short History of the Auxiliary Infantry Units of the Imperial Roman Army*. BAR International Series 841. Oxford: Archaeopress.

Speidel, M. Alexander. 1992. "Roman Army Pay Scales." *Journal of Roman Studies* 82: 87–106.

———. 2014. "Roman Army Pay Scales Revisited: Responses and Answers." Pages 53–62 in *De l'or pour les Braves! Soldes, Armées et Circulation Monétaire dans le Monde Romain*. Edited by Michel Reddé. Collection Scripta Antiqua 69. Bordeaux: Ausonius Éditions.

Speidel, Michael P. 1973. "The Pay of the Auxilia." *Journal of Roman Studies* 63: 141–147.

———. 1982–1983. "The Roman Army in Judaea under the Procurators: The Italian and the Augustan Cohort in the Acts of the Apostles." *Ancient Society* 13–14: 233–240.

———. 1992. "Becoming a Centurion in Africa: Brave Deeds and Support of the Troops as Promotion Criteria." Pages 124–128 in *Roman Army Studies: Volume Two*. Edited by Michael P. Speidel. Mavors Roman Army Researches 8. Stuttgart: Steiner.

Sperber, Daniel. 1965. "The Costs of Living in Roman Palestine I." *Journal for the Economic and Social History of the Orient* 8: 248–271.

———. 1966. "The Costs of Living in Roman Palestine II." *Journal for the Economic and Social History of the Orient* 9: 182–211.

———. 1991. *Roman Palestine 200–400: Money and Prices*. 2nd ed. Bar-Ilan Studies in Near Eastern Languages and Culture. Ramat Gan: Bar-Ilan University Press.

Standhartinger, Angela. 2015. "Letter from Prison as Hidden Transcript: What It Tells Us about the People at Philippi." Pages 107–140 in *The People beside Paul: The Philippian Assembly and History from Below*. Edited by Joseph A. Marchal. Early Christianity and Its Literature 17. Atlanta: SBL.

Stegemann, Ekkehard W. and Wolfgang Stegemann. 1999. *The Jesus Movement: A Social History of Its First Century*. Translated by O. C. Dean, Jr. Minneapolis: Fortress.

Stegemann, Wolfgang. 1987. "War der Apostel Paulus ein römischer Bürger?" *Zeitschrift für die Neutestamentliche Wissenschaft* 78: 200–229.

Stiebel, Guy D. 2015. "Military Dress as an Ideological Marker in Roman Palestine." Pages 153–167 in *Dress and Ideology: Fashioning Identity from Antiquity to the Present*. Edited by Shoshana-Rose Marzel and Guy D. Stiebel. London: Bloomsbury.

Stoll, Oliver. 2001. *Zwischen Integration und Abgrenzung: Die Religion des Römischen Heeres im Nahen Osten. Studien zum Verhältnis von Armee und Zivilbevölkerung im römischen Syrien und in Nachbargebieten.* Mainzer Althistorische Studien 3. St. Katharinen: Scripta Mercaturae.

———. 2007. "The Religions of the Armies." Pages 451–476 in *A Companion to the Roman Army.* Edited by Paul Erdkamp. Blackwell Companions to the Ancient World. London: Blackwell.

Stratton, Kimberly B. 2009. "The Eschatological Arena: Reinscribing Roman Violence in Fantasies of the End Times." *Biblical Interpretation* 17: 45–76.

Such, W. A. 1999. *The Abomination of Desolation in the Gospel of Mark: Its Historical Reference in Mark 13:14 and Its Impact in the Gospel.* Lanham: University Press of America.

Sumner, Graham. 2002. *Roman Military Clothing (1): 100 BC–AD 200.* Men-at-Arms 374. Oxford: Osprey.

Syon, Danny. 2015. *Small Change in Hellenistic-Roman Galilee: The Evidence from Numismatic Site Finds as a Tool for Historical Reconstruction.* Numismatic Studies and Researches 11. Jerusalem: Israel Numismatic Society.

Tajra, Harry W. 2010. *The Trial of St. Paul: A Juridical Exegesis of the Second Half of the Acts of the Apostles.* Eugene: Wipf and Stock.

Thompson, Leonard L. 1990. *The Book of Revelation: Apocalypse and Empire.* Oxford: Oxford University Press.

Toner, Jerry P. 2002. *Rethinking Roman History.* Cambridge: Oleander.

Trebilco, Paul. 1991. *Jewish Communities in Asia Minor.* Society for New Testament Studies Monograph Series 69. Cambridge: Cambridge University Press.

Tsafrir, Yoram. 1982. "The Desert Fortresses of Judaea in the Second Temple Period." Pages 120–145 in *The Jerusalem Cathedra: Studies in the History, Archaeology, Geography and Ethnography of the Land of Israel.* Edited by Lee I. Levine. Vol. 2 of 3. Detroit: Wayne State University Press.

Udoh, Fabian E. 2005. *To Caesar What Is Caesar's: Tribute, Taxes, and Imperial Administration in Early Roman Palestine 63 BCE–70 CE.* Brown Judaic Studies 343. Providence: Brown Judaic Studies.

Vaage, Leif E. 2006. "Why Christianity Succeeded (in) the Roman Empire." Pages 253–278 in *Religious Rivalries in the Early Roman Empire and the Rise of Christianity.* Edited by Leif E. Vaage. Studies in Christianity and Judaism 18. Waterloo: Wilfred Laurier University Press.

———. 2014. *Borderline Exegesis.* Signifying (on) Scriptures. University Park: Pennsylvania State University Press.

Wansink, Craig S. 1996. *Chained in Christ: The Experience and Rhetoric of Paul's Imprisonments.* Journal for the Study of the New Testament: Supplement Series 130. Sheffield: Sheffield Academic.

Ward, Graeme A. 2012. *Centurions: The Practice of Roman Officership.* Ph.D. Dissertation, Classics, University of North Carolina at Chapel Hill.

Weaver, Dorothy Jean. 2005. "'Thus You Will Know Them by Their Fruits': The Roman Characters of the Gospel of Matthew." Pages 107–127 in *The Gospel of Matthew in Its Roman Imperial Context.* Edited by John Riches and David C. Sim. Journal for the Study of the New Testament: Supplement Series 276. London: T&T Clark International.

Webber, Christopher. 2001. *The Thracians 700 BC–AD 46.* Men-at-Arms 360. London: Osprey.

Weiss, Zeev. 2014. *Public Spectacles in Roman and Late Antique Palestine.* Revealing Antiquity 21. Cambridge: Harvard University Press.

Wendt, Heidi. 2015. "*Ea Superstitione*: Christian Martyrdom and the Religion of Freelance Experts." *Journal of Roman Studies* 105: 183–202.

Wengst, Klaus. 1987. *Pax Romana and the Peace of Jesus Christ*. Translated by John Bowden. London: SCM Press.

Wheeler, Everett. 1993. "Methodological Limits and the Mirage of Roman Strategy." *Journal of Military History* 57: 7–41, 215–240.

———. 2007. "The Army and the *Limes* in the East." Pages 235–266 in *A Companion to the Roman Army*. Edited by Paul Erdkamp. Blackwell Companions to the Ancient World. London: Blackwell.

Whitehorne, John. 2004. "Petitions to the Centurion: A Question of Locality?" *Bulletin of the American Society of Papyrologists* 41: 155–169.

Whittaker, C. R. 1994. *Frontiers of the Roman Empire: A Social and Economic Study*. Ancient Society and History. London: Johns Hopkins University Press.

Wifstrand, Albert. 1939. "Autokrator, Kaisar, Basileus." Pages 529–539 in *ΔΡΑΓΜΑ: Martino P. Nilsson, A.D. IV id. iul. anno MCMXXXIX dedicatum*. Edited by H. Ohlsson. Skrifter Utgivna av Svenska Institutet i Rom 1. Lund: Ohlssons.

Wigg, David G. 1997. "Coin Supply and the Roman Army." Pages 281–288 in *Roman Frontier Studies 1995*. Edited by W. Groenman-van Waateringe, B. L. van Beek, W. J. H. Willems, and S. L. Wyrria. Oxbow Monograph Series 91. Oxford: Oxbow.

Wilmanns, Juliane C. 1995. *Der Sanitätsdienst im römischen Reich: Eine sozialgeschichtliche Studie zum römischen Militärsanitätswesen neben einer Prosopographie des Sanitätspersonals*. Medizin der Antike 2. Hildesheim: Olms-Weidmann.

Woolf, Greg. 1998. *Becoming Roman: The Origins of Provincial Civilization in Gaul*. Cambridge: Cambridge University Press.

Wright, N. T. 2009. *Paul: In Fresh Perspective*. Minneapolis: Fortress.

———. 2011. "Paul and Empire." Pages 285–297 in *The Blackwell Companion to Paul*. Edited by Stephen Westerholm. Oxford: Wiley-Blackwell.

Yoder, John Howard. 1996. "Is There Such a Thing as Being Ready for Another Millennium?" Pages 63–69 in *The Future of Theology: Essays in Honor of Jürgen Moltmann*. Edited by Miroslav Volf, Carmen Krieg, and Thomas Kucharz. Grand Rapids: Eerdmans.

Yoder, Joshua. 2014. *Representatives of Roman Rule: Roman Provincial Governors in Luke-Acts*. Beihefte zur Zeitschrift für die neutestamentliche Wissenschaft 209. Berlin: De Gruyter.

Yoder Neufeld, Thomas R. 1997. *"Put on the Armour of God": The Divine Warrior from Isaiah to Ephesians*. Journal for the Study of the New Testament: Supplement Series 140. Sheffield: Sheffield University Press.

Zeichmann, Christopher B. 2012. "οἱ στρατηγοὶ τοῦ ἱεροῦ and the Location of Luke–Acts' Composition." *Early Christianity* 3: 172–187.

———. 2015. "Herodian Kings and Their Soldiers in the Acts of the Apostles: A Response to Craig Keener." *Journal of Greco-Roman Christianity and Judaism* 11: 178–190.

———. 2015. "Rethinking the Gay Centurion: Sexual Exceptionalism, National Exceptionalism in Readings of Matt 8:5–13/Luke 7:1–10." *Bible and Critical Theory* 11: 35–54.

———. 2017. "Capernaum: A 'Hub' for the Historical Jesus or the Markan Evangelist?" *Journal for the Study of the Historical Jesus* 15: 147–165.

———. 2017. "The Date of Mark's Gospel Apart from the Temple and Rumors of War: The Taxation Episode (12:13–17) as Evidence." *Catholic Biblical Quarterly* 79: 422–437.

———. 2017. "Loanwords or Code-Switching? Latin Transliteration and the Setting of Mark's Composition." *Journal of the Jesus Movement in Its Jewish Setting* 4: 42–64.

———. 2018. "Gender Minorities In and Under Roman Power: Respectability Politics in Luke–Acts." Pages 61–73 in *Luke-Acts*. Edited by James Grimshaw. Texts@Contexts. London: Bloomsbury.

———. 2018. "Liberal Hermeneutics of the Spectacular in the Study of the New Testament and the Roman Empire." *Method and Theory in the Study of Religion* 30: forthcoming.

———. 2018. "Military Forces in Judaea 6–130 CE: The *status quastionis* and Relevance for New Testament Studies." *Currents in Biblical Research* 17: forthcoming.

Zissu, Boaz and Avner Ecker. 2014. "A Roman Military Fort North of Bet Guvrin/Eleutheropolis." *Zeitschrift für Papyrologie und Epigraphik* 188: 293–312.

Žižek, Slavoj. 2008. *Violence: Six Sideways Reflections*. Big Ideas//Small Books. New York: Picador.

———. 2009. *First as Tragedy, Then as Farce*. London: Verso.

Index

Aelia Capitolina. *See* Jerusalem

Agrippa I, 4, 5, 14, 17n15, 32–33, 66, 79, 83–86, 91

Agrippa II, 4, 5, 14, 17n15, 26, 33, 89–91, 105n81

ala Sebastenorum, 4, 17n6, 17n8

ala/alae, 2, 4, 52, 136n6

ambivalence, xviii–xix, 53, 60, 65, 75, 135, 139–143

Antipas, Herod, 2, 4, 5, 18n23, 32, 46n47, 57–58, 67–70, 81–82, 91–92

angels, 52, 56–57, 71–72, 75, 86, 109–110, 128–130

aqueduct construction, 34–35. *See also* infrastructure, construction of

Archelaus, Herod, 4, 32, 66, 79, 90, 105n89

Ascalon, xix, 17n7, 27, 45n43, 78

Augustan cohort. *See cohors Augusta*

Augustus, emperor, 101n31, 134

auxilia, 2–16, 24–25, 32–33, 39, 52, 83–91. *See also ala, cohors, cohors equitata*

bandits, 4, 25, 26–27, 35, 60, 72, 81, 127

Bar Kokhba War, 5, 26, 33, 36, 40–41. *See also* Hadrian, emperor

Batanaea: royal army, 5, 7, 27, 32–33, 45n43, 83–86, 89–91. *See also* Agrippa II; Philip, Herod

bodyguarding, 1, 24, 26, 44n30, 116–117

Caesarea Maritima, 3–8, 34–35, 38, 83–85, 88–91

Caligula, emperor. *See* Gaius, emperor

Caparcotna/Legio, 31, 38

Capernaum, 31–33, 50, 67–70, 78–79, 93–94

centurions, xvii, 5, 6, 8, 9, 13–15, 27–29, 31, 38–40, 60–65, 67–70, 74, 78–79, 83–85, 89–91, 93; recruitment and promotion, 13–15

Christian soldiers, 84, 143n2

Claudius, emperor, 69, 87, 132

clothing, military, 8–9, 39, 72–73

cohors/cohortes, 2, 4, 52

cohors Augusta, 5, 89–91

cohors II Italica civium Romanorum, 83–85

cohortes Sebastenorum, 4, 17n6, 17n8

cohors equitata/cohortes equitatae, 2, 88–89

coinage, 36–38, 120. *See also* countermarks

colonies, military, 4–5, 7, 11, 27, 39

Constantine, emperor, 84–85, 140, 143n2

countermarks, 36, 46n45, 50
crucifixion, 60–65, 82–83, 96–98

Decapolis: army of, 30–31, 34, 50–57;
    soldiers from, 6, 17n13. *See also* Gerasa
diplomas, 4, 6, 10–11, 20n33, 21n43
disciplina militaris, 18n19, 122
Domitian, emperor, 100n27, 111, 125,
    130–131

economy, military and, 8–11, 30, 34, 35–
    40, 44–45n32, 46n47, 133, 140–141.
    See also coinage; countermarks
Egypt, 2, 6, 24, 27–29, 40, 135
emperor: military role, 88, 90, 109–112,
    130–131; titles of, 6, 60–65, 95–96,
    100–101n27, 101n33, 101n35,
    128–129
extortion, 28–29, 35, 36, 67, 76–77

families of soldiers, 7, 40, 113
food, 8–10, 27, 39, 113–114
fortifications/fortlets/watchtowers, 29–31,
    43–44n25

Galilee: royal army, 32–33, 45n43, 57–58,
    67–70. *See also* Antipas, Herod
Gaius, emperor, 28, 117
Gerasa, 38, 40, 50–57, 62, 79
grand strategy, Roman, 24–25

Hadrian, emperor, 34–35, 51, 87, 122
Hasmonaeans, 4, 5, 54, 111
Herod the Great, xviii, xix, 3–5, 8, 10,
    43–44n25, 65–66, 87

infrastructure, construction of, 33–35,
    44–45n25, 46n47
Israel-Palestine, xix–xx, 25
Italian cohort. *See cohors II Italica civium*
    Romanorum

Jerusalem, 5, 28, 31, 34, 36–38, 51–52,
    57, 79–80, 86–88. *See also* Temple,
    destruction of
Jewish War, differences before and after,
    1–2, 12–16, 29–33, 36–41. *See also*
    Temple, destruction of

John the Baptist, 29, 57–58, 70, 76–78
Judaea: annexation of, 4, 12, 16n4, 32–33;
    Judaean soldiers, 3–6. *See also* Agrippa I;
    Archelaus, Herod; Herod the Great
Judaism/Jewishness: authenticity politics and
    definition of, xx, 6–7, 17n12, 17n16;
    compatibility with military service, 6–7,
    10, 60–65; Jewish soldiers, 5–6

language of the military, 2, 12–13, 50
legio X Fretensis, 12–13, 31, 37, 40, 46n45,
    50–57
legions/legionaries, 1–3; Judaean legions,
    12–13, 32–33, 40–41, 50–57
loyalty, 2, 15, 54, 62–63, 77, 95, 116–117,
    121–122

Nero, emperor, 69, 90, 117, 125

Palestine, definition of, xix–xx, 30
Palestinian soldiers, 5–6
parousia, 109–112
Parthia, 12, 14, 21n48, 24–25, 127, 136n5
Paul, 86–91, 107–124; Acts' depiction
    vs. epistles, 86, 91, 108; authentic vs.
    disputed epistles, 107, 109, 118, 120
persecution of Christians, 108, 115, 127,
    130–131, 136n13
Philip, Herod, 4, 5, 32
pigs, 10 50–57
Pontius Pilate, 26, 34, 60–61, 73–75,
    81–82, 95, 99n21
praetorian guard, 2–3, 116–117
prison, 85–86, 114–117
prisoners of war, 112, 114–116

recruitment, 3–7, 127
religion in the army, 6–7, 10, 17–18n17,
    60–65
Roman citizenship, 1–2, 4, 10–12, 39,
    84, 86–87, 108. *See also* diplomas; *tria*
    nomina
royal armies, 2–7

sex and soldiers, 10, 27, 67–70, 102–
    103n46, 103n49, 132–133
slaves, 36, 40, 67–71, 78–80, 84, 87, 97,
    110, 114–115, 117

supersessionism, 79–80, 91, 118
Syria Palaestina. *See* Judaea

tax collectors, 31, 34, 37–38, 46n47, 76
Temple, destruction of, 3, 6, 13, 23, 37, 50, 58–59, 70–71, 79–80, 111, 125, 133, 139
Tiberius Julius Alexander, xx, 6, 14–15, 21n49
Tiberius, emperor, 8
Titus, emperor, 13–14, 58–59, 111. *See also* Temple, destruction of
Trajan, emperor, 12, 14, 36, 43–44n25, 45n43, 78
tria nomina, 6, 17n14, 121–122. *See also* Roman citizenship
triumph, 109–112, 133–135

Vespasian, emperor, 28, 37, 43–44n25, 100–101n27, 125

wages of soldiers, 8–9, 29, 39, 76–78, 113–114
War of Quietus, 76, 77
warfare: siege combat, 14, 59, 62, 114–115; spiritual warfare, 49, 72, 112–113, 118–120, 129–130. *See also* Bar Kokhba War; Jewish War, differences before and after; War of Quietus
weapons and armor, 71–72, 79–81, 112–113, 118–120, 126–127, 135
women. *See* families of soldiers; sex and soldiers; slaves

Author Index

Aldrete, Alicia, 137n23
Aldrete, Gregory S., 137n23
Alföldy, Géza, 19n26
Allison, Dale C., Jr., 101n31
Alston, Richard, 41n2, 43n15, 43n16, 43n23
Alt, Albrecht, 43–44n25
Álvarez Cineira, David, 101n32
Amit, David, 45n34
Ando, Clifford, 124n33
Applebaum, Shimon, 17n9, 43n19, 47n52, 105n82
Arnal, William E., 34, 44–45n32, 45n33, 46n47, 47n53, 50, 54–56, 98n2, 99n12, 99n14, 104n65
Aubert, Jean-Jacques, 43n15
Aune, David E. 131–132, 136n2, 136n10, 137n17
Avalos, Hector, 140, 143n3
Aviam, Mordechai, 44n31

Baker, Patricia Anne, 19n31
Ball, Warwick, 41n4
Barclay, John M. G., 21n50
Bartell, Scott M., 137n23
Bauckham, Richard, 137n18
Baxter, Fred, 7
Bazzana, Giovanni B., 106n97
Beale, G. K., 128, 136n2, 136n7, 136n9

Bermejo-Rubio, Fernando, 99n21
Bernett, Monika, 101n33
Bingham, Sandra, 124n23
Birley, Anthony R., 106n94
Birley, Eric, 15, 20n33, 21n51, 21n54, 105n78
Bligh, Philip H., 63, 101n32, 101n33
Bourdieu, Pierre, 143n1
Bowens, Lisa M., 123n15
Braund, David, 106n95
Breeze, David., 21n52
Brent, Allen, 100n27
Breytenbach, Cilliers, 123n12
Brink, Laurena Ann, xxin5, 65, 67, 83, 102n44, 104n69, 105n87
Brown, Raymond E., 97, 101n30, 102n39, 106n99

Campbell, R. Alastair, 134, 137n22
Caragounis, Chrys C., 123n16
Carreras Monfort, César, 45n41
Carter, Warren, 98n3, 99n5
Cassidy, Richard J., 104n73
Catchpole, David R., 110–111, 123n9, 123n10
Chancey, Mark A., 46n47, 48n70
Charlesworth, James H., 44n31
Christian, Michelle, 106n98
Christiansen, Erik, 46n48

169

Cloud, Dana L., 144n5
Cohen, Getzel M., 17n9
Collins, Adela Yarbro, 123n10, 131, 137n14
Cook, John Granger, 100n23, 101n35
Cotton, Hannah M., 41n4, 47n54, 47n57, 47n63
Crawford, Michael, 46n48
Crossan, John Dominic, 41n1, 122n6
Cumont, Franz, 123n21

Davies, Gwyn, 47n54
Davies, Jeffrey L., 37, 45n41, 47n51
Davies, Roy W., 17n14, 19n28, 19n29, 19n31, 43n15, 47n61, 123n17
de la Bédoyère, Guy, 124n23
Devijver, Hubert, 87–88, 105n77, 105n78
Di Segni, Leah, 18n20, 45n36
Dobson, Brian, 21n51
Donahue, John R., 52, 99n6
Downing, Francis Gerald, 103n52
Drinkwater, John, 16n4
Droge, Arthur J., 124n31
Duff, Paul Brooks, 123n10
Dyson, Stephen, 41n4

Eck, Werner, 45n43
Ecker, Avner, 44n28, 78, 103n62
el-Khouri, Lamia, 44n27
Elliott, Neil, 112, 123n11, 123n12
Elton, Hugh, 41n4
Emerson, Ralph Waldo, 140–141, 144n6
Engberg-Pedersen, Troels, 99n9
Eubank, Nathan, 100n24

Ferrill, Arther, 41n4
Février, P. A., 45n35
Fewster, Gregory P., 122n4
Fiensy, David A., 19n26
Fink, Robert O., 17n14, 48n68
Fischer, Moshe, 43n24, 44n26, 63n47
Fitzmyer, Joseph A., 104n64
Flexsenhar, Michael A., III, 124n24
Forni, Giovanni, 20n38
Fortna, Robert T., 106n91
France, R. T., 65–66, 40n40
Fredriksen, Paula, 17n12, 103n52
Freeman, Philip, 119–120, 124n29

Friedheim, Emmanuel, 100–101n27
Friesen, Steven J., 105n85, 122n2, 136n12
Fuhrmann, Christopher J., 42n12
Fusco, Vittorio, 104n64

Garnsey, Peter D., 9, 18n25
Gelardini, Gabriella, xxin5
Gichon, Mordechai, 25, 42n8, 43–44n25
Gilliver, Kate, 21n51
Glancy, Jennifer A., 132–133, 137n20
Goldsmith, Raymond W., 19n27
Gombis, Timothy G., 124n28
González Salinero, Raúl, 17n11
Goodacre, Mark, 80
Gracey, M. H., 43–44n25
Graf, D. F., 16n4, 21n42
Green, Joel B., 104n64
Groenman-van Waateringe, W., 19n28
Groh, Dennis E., 103–104n63
Guthrie, George H., 123n18

Haensch, Rudolf, 18n17, 100n26
Halperin, David M., 132, 137n19
Hamel, Gildas, 19n26
Hanson, A. E., 44n30, 47n58
Har-Peled, Misgav, 57, 99n16
Harrington, Daniel J., 52, 99n6
Haynes, Ian, 16n4, 47n60
Head, Duncan, 136n3
Heemstra, Marius, 137n16
Heilig, Christopher, 123n12
Hekster, Olivier J., 100n27
Hobbs, T. R., 48n71
Hoey, Allan S., 17n17, 100n26
Hölbl, Günther, 100n27
Holder, Paul A., 20n33
Holloway, John, 143, 144n11
Horsley, Richard A., xviii, xxin2, 41n1, 53–54, 66, 98n3, 99n5, 99n8, 99n9, 100n24, 102n42
Howgego, Christopher, 46n45

Incigneri, Brian J., 59, 98n3, 99n20
Isaac, Benjamin, 6, 17n15, 25, 27, 31, 41n2, 41n4, 42n5, 42–43n13, 43n14, 43n22, 43n24, 44n26, 44n29, 44n31, 44–45n32, 45n43, 122n5
Iverson, Kelly R., 100n24

James, Simon, 91–92, 105–106n90
Jennings, Theodore W., Jr., 68–69, 103n47
Jensen, Morten Hørning, 46n47
Johnson, Brian, 97, 106n102
Johnson, Earl S., 61–62, 100n24, 100n25
Judge, E. A., 106n96
Juel, Donald, 100n24

Kagan, Kimberly, 41–42n4
Kalantzis, George, 104n71
Kasher, Aryeh, 16n3
Katsari, Constantina, 46n47
Kelber, Werner H., 135, 137n24
Kennedy, David L., 31, 42n4, 44n27
Kilpatrick, G. D., 97, 106n100, 122n8
Kim, Seyoon, 53, 99n7
King, Anthony C., 19n28
Kittel, Gerhard, 123n20
Klauck, Hans-Josef, 89, 105n80
Klinghardt, Matthias, 98n3
Kloppenborg, John S., 58, 72, 99n18, 101n29, 103n53, 104n68
Koester, Craig R., 127, 136n5, 136n10, 137n18
Koester, Helmut, 122n6
Koskenniemi, Erkki, 102n41
Krause, Jens-Uwe, 104n73, 104n75
Kreissig, Heinz, 19n26
Kyrychenko, Alexander, 91, 105n87, 105n88, xxin5

Lambrecht, Jan, 103n58
Lapin, Hayim, 47n49
Lau, Markus, 98n3
Leander, Hans, 98n3, 99n10, 123n10
Lewis, Naphtali, 47n56
Liew, Tat-Siong Benny, 68–69, 103n47
Lincoln, Bruce, 143, 144n12
Llewelyn, S. R., 21n47
Long, Fredrick J., 124n28
Luttwak, Edward N., 24–25, 41n3, 41–42n4, 42n7

MacDonald, Dennis R., 124n32
Mack, Burton L., 106n91
Macky, Peter W., 122n1
MacMullen, Ramsay, 41n2, 45n36
Magness, Jodi, 20n39, 47n54

Maier, Harry O., 124n27
Malherbe, Abraham J., 114, 123n18
Malina, Bruce J., 128–129, 136n8
Mann, J. C., 20n33, 20n38, 42n4, 48n69
Martin, Dale B., 71, 103n51, 103n52
Martin, Luther H., 99n14, 130, 136n11
Mason, Hugh J., 20–21n42
Mason, Steve, xxin6, 14, 16n1, 21n44, 21n46
Mattern, Susan P., 42n4
Mattingly, David J., 103n49
McCutcheon, Russell T., 54–56, 99n12, 99n14
Meggitt, Justin J., 18n25, 19n26
Mélèze-Modrzejewski, Joseph, 21n49
Metzger, Bruce M., 97, 106n101
Michalak, Aleksander R., 99n13
Millar, Fergus, 42n4, 122n5
Miller, Amanda C., 103n57
Mitchell, Stephen, 101–102n36
Moore, Stephen D., 132–133, 137n21
Mounce, Robert H., 136n4

Nicolotti, Andrea, 100n22
Noethlichs, Karl-Leo, 104–105n76
Nolland, John, 104n64
Nossov, Konstantin, 123n14

Osborne, Grant R., 136n4
Oudshoorn, Jacobine G., 47n55, 47n56

Patrich, Joseph, 45n34
Pažout, Adam, 44n25
Peachin, Michael, 43n15, 43n18
Perkins, Pheme, 56, 99n15
Pervo, Richard I., 91, 103n58, 105n86, 122n3
Peters, Edward M., 104n72
Phang, Sara Elise, 11, 20n34, 20n36, 48n67m 124n33
Pilch, John J., 128–129, 136n8
Pollard, Nigel, 17n14, 20n38, 20n40, 39, 41n2, 42n5, 44n31, 46n47, 48n65, 48n67, 48n71
Price, S. R. F., 101n35
Pucci Ben Zeev, Miriam, 36, 45n42

Rapske, Brian, 104n73
Reed, Jonathan L., 44n31, 122n6

Richardson, Peter, 42n11, 102n43
Richmond, I. A., 19n32
Rocca, Samuel, xxin3, 4, 16n3, 17n11
Roll, Israel, 43n24, 44n26
Rollens, Sarah E., 19n26
Roller, Duane W., 43n22
Rosenberger, Mayer, 46n45
Rosenthal-Heginbottom, Renate, 45n37
Rossing, Barbara, 137n20
Roth, Jonathan P., 4, 16n1, 16n2, 17n11,
    19n30, 37, 39, 45n33, 45n41, 45n43,
    46n48, 47n50, 47n50, 47n62, 76,
    103n60
Roxan, Margaret M., 20n35
Russell, James, 20n37

Saddington, Denis B., xviii, xxin4, 69–70,
    103n48, 103n61
Safrai, Ze'ev, 43n19, 45n33, 45n41, 47n50,
    48n70
Sanders, E. P., xvii–xviii, xxn1
Sartre, Maurice, 42n4, 122n5
Scheidel, Walter, 18n25
Schellenberg, Ryan S., 122n3
Schoenfeld, Andrew J., 7, 17n11, 18n18
Schulten, Adolf., 19n32
Schultz, Brian, 105n89
Schürer, Emil, 16n5, 99n11, 99n17,
    104n70
Schwartz, Daniel R., 102n41
Schwartz, Saundra, 86, 104n74
Schwartz, Seth, 76, 103n59
Segal, Arthur, 47n59
Shatzman, Israel, xxin3, 16n3, 17n10,
    18n23, 29–30, 43n25
Shaw, Brent D., 136n13
Shea, Christine R., 105n83
Shean, John F., 104n71
Sherwin-White, Adrian N., 104n67
Shiner, Whitney T., 63–64, 101n30,
    101n34
Shotwell, Alexis, 141–143, 144n7, 144n11
Sider, Ronald J., 104n71
Smallwood, E. Mary, 99n11
Smith, Dennis E., 105n84
Smith, Julien, 124n26
Spaul, John, 21n43
Speidel, M. Alexander, 18n21

Speidel, Michael P., 18n21, 21n51, 21n53,
    83–84, 90, 104n70, 105n81
Sperber, Daniel, 18n22
Standhartinger, Angela, 124n24
Stegemann, Ekkhard W., 105n76
Stegemann, Wolfgang, 104–105n76
Stiebel, Guy D., 100–101n27
Stoll, Oliver, 17–18n17, 47n50, 62,
    100n26, 101n28
Stratton, Kimberly B., 135–136, 137n25
Such, W. A. 59, 99n19
Sumner, Graham, 72–73, 103n54, 103n55
Syon, Danny, 36–37, 46n46, 46n48

Tajra, Harry W., 105n76
Thompson, Leonard L., 131, 137n15
Toner, Jerry P., 19n26
Trebilco, Paul, 101–102n36, 102n37
Tsafrir, Yoram, 44n25
Tyson, Joseph B., 105n84

Udoh, Fabian E., 46n44

Vaage, Leif E., 125, 136n1, 144n10

Wansink, Craig S., 104n73, 123n20,
    124n22
Ward, Graeme A., 48n66
Weaver, Dorothy Jean, 103n56
Webber, Christopher, 126, 136n3
Weiss, Zeev, 47n59
Weksler-Bdolah, Shlomit, 18n20
Wendt, Heidi, 136n13
Wengst, Klaus, 122n7
Whitehorne, John, 43n15, 43n18
Whittaker, C. R., 42n4, 42n5, 103n49
Wifstrand, Albert, 96, 106n96
Wigg, David G., 46n47
Wilmanns, Juliane C., 19n31
Woolf, Greg, 9, 18n25, 39, 47n64
Wright, N. T., 41n1, 124n28

Yoder, John Howard, 143n2
Yoder, Joshua, 105n79
Yoder Neufeld, Thomas R., 123n13

Zissu, Boaz, 44n28, 78, 103n62
Žižek, Slavoj, 140–142, 143n4, 144n8

Index of the Bible and Other Ancient Texts

Hebrew Bible

Genesis
  14:18–20 (LXX), 102n37

Exodus
  12:46, 96

Numbers
  24:16 (LXX), 102n37

Joshua
  10, 56

Esther
  8E:17 (LXX-OG), 102n37

Job
  39:19–25, 128

Psalms
  2:1–11, 134
  9:6 (Vulgate), 126
  18:8–16, 56
  22, 64
  22:1, 64
  22:19, 60

34:20, 96
44:22, 115
45:3–4, 134
69:21, 60
137:9, 66

Isaiah
  6:9–10, 63
  11:4, 126
  34:12, 128
  49:2, 126
  53:12, 80
  59:17, 112
  63:1–6, 134

Jeremiah
  31:15, 66

Daniel
  3:93 (LXX-TH, LXX-OG), 102n37
  4:2 (LXX-TH), 102n37
  4:17 (LXX-TH), 102n37
  4:34 (LXX-TH, LXX-OG), 102n37
  4:37 (LXX-OG), 102n37

5:21 (LXX-TH), 102n37
9:27 (LXX-TH, LXX-OG), 59
10:10–13, 56

Joel
  1–2, 128
  1:16, 128
  2:4–7, 128
  3:11–13, 134

Amos
  8:9, 101n31

Deuterocanonical Literature

1 Maccabees
  1:54–56, 59
  3:28, 77
  13:49–51, 111

2 Maccabees
  3:31, 102n37
  4:21–22, 111
  8:12, 110

15:21, 110
15:22–23, 56

3 Maccabees
7:9, 102n37

1 Esdras
2:2, 102n37
6:30, 102n37
8:19, 102n37
8:21, 102n37

2 Esdras
11:1, 59

2 Baruch
6, 129

3 Baruch
4:7, 56

Wisdom of Solomon
5:17–20, 112
18:15–16, 126

New Testament

Matthew
1:1–17, 134
2:13–16, 65–66
5:38–42, 67, 71
5:41, 67
8:5–13, xxin4, 18,
   20n42, 31, 67–70,
   75, 78–79, 93
8:6, 68
8:8, 68
8:9, 68
8:11–12, 68
8:28–34, 75
10:18, 70
10:34–36, 71
14:1–12, 70
15:8, 36
17:24, 36
20:1–15, 8

24:1–51, 70–71
26:52, 126
26:52–54, 71–72
26:53, 56
27:25, 72
27:27, 74
27:27–31, 72–74
27:34, 96
27:44, 72
27:54, 74
27:54–28:15, 74–75
28:18–20, 68

Mark
2:5–12, 62
2:7, 55
2:13–14, 31
3:2–6, 62
3:20–22, 62
3:22, 57
3:22–27, 72
4:12, 63
5:1–20, xvii, 50–57
5:7, 64
5:9, 50, 52
5:11, 99n5
5:13, 99n5
5:15, 50
5:16, 55
5:17, 53
5:20, 53, 55
6:14–29, 57–58
6:16, 58
6:17, 58
6:21, 57, 73
6:26, 57
6:27–28, 57
7:34, 64
8:27–29, 55
8:38, 56, 72
9:12–13, 58
10:17–22, 64
10:17–31, 73
11:1–11, 111
12:5, 36
12:42, 36
13:1–2, 58–59

13:1–37, 58–59, 133
13:2, 58
13:7–8, 59
13:9, 70
13:14, 59
13:14–18, 59
13:24–27, 56, 72
14:10–11, 73
14:43, 81
14:47, 71
15:1–15, 60
15:2, 60
15:6, 64
15:15, 6
15:16, 116
15:16–20, 72–74
15:16–32, xviii, 60
15:20, 62
15:21–32, 60
15:23, 60, 96
15:24, 60
15:26, 60
15:29–30, 61
15:31–32, 61
15:32, 61
15:34, 64
15:35, 64
15:39, 60–65
15:44, 60–65
15:44–45, 61, 63

Luke
2:2, 75
2:13–15, 56
2:35 (Syriac), 126
3:1–2, 75
3:14, xvii, 18n23, 29,
   76–78, 91
3:19–20, 78
7:1–10, xvii, 20n42, 31,
   67–70, 77, 78–79,
   93
8:26–39, 79
10:3, 81
10:4, 80
13:1–5, 34
13:35, 58

19:11–27, 105n89
21:5–38, 79–80
21:20, 79
21:24, 79, 123n20
22:4, 81
22:35–38, 80–81
22:35–53, 80–81
22:52, 81
23:1–25, 81–82
23:35–48, 82–83

John
2:13, 94
4:45, 94, 106n92
4:45–54, 93–94
4:46, 106n92
4:47, 106n92
4:49–52, 106n92
6:64–71, 96
8:44, 96
13:2, 96
18:1–12, 94–95
18:6, 94
18:11, 94
18:28–19:22, 95–96
19:1, 42n10
19:23–25, 96
19:23–37, 96–98
19:29, 97
19:33–37, 96

Acts
3:14, 83
4–5, 81
5:33–39, 91
7:2–53, 80
8:26–40, 88
9:32–35, 90
10:1, 83–84
10:1–48, 83–85
12:1–19, 85–86
16:17, 102n37
16:22–23, 42n10
21:17–22:30, 86–88
21:38, 88, 91
22:3, 108
22:28, 87

23:1–35, 88–89
23:35, 116
27:1–28:31, 89–91

Romans
6:13, 112–113
7:23, 107
8:31–39, 115
8:37, 114–115
8:37–39, 107
13:1–7, 107
13:12, 112–113, 118
16:7, 115–116

1 Corinthians
2:8, 107
7, 121
7:8, 120
7:25–28, 120
9:7, 77, 107, 113–114
9:25, 107
14:8, 110
15:20–28, 109–112
15:24, 112
15:27, 112
15:52, 109–112

2 Corinthians
2:14, 111, 115–116
2:14–16, 109–112
4–6, 123n15
6:7, 112–113, 118
10:1–13:10, 114
10:3–6, 114–115
10:5, 123n20
11:6, 108, 116
11:12, 123n19
11:16–33, 117, 121
11:24–25, 42n10

Galatians
2:1–10, 109

Ephesians
1:2, 119
1:20, 119
1:20–23, 119

1:21, 119
2:4, 119
2:11–22, 119
2:14, 119
2:17, 119
3:14, 119
3:17, 119
4:2, 119
4:15–16, 119
4:26, 119
4:32, 119
5:2, 119
5:5. 119
5:21–6:9, 124n30
5:28, 119
5:33, 119
6:10–17, 118–120
6:12, 119
6:17, 126
6:20, 119
6:23–24, 119

Philippians
1:1–3:1, 116
1:10, 118
1:12–13, 116–117
1:21–23, 118
2:25, 115–116
3:20–4:1, 118
4:22, 117

Colossians
1:29, 107
2:15, 123n12
4:10, 116

1 Thessalonians
1:9–10, 112
2:2, 107
4:12, 123n11
4:16–17, 109–112
5, 123n13
5:1, 110
5:8, 112–113

1 Timothy
1:18, 121–122

2 Timothy
2:2, 121
2:3–4, 121–122

Philemon
2, 115–116
23, 115–116

Hebrews
4:12, 126

James
4:1, 107

1 Peter
2:11, 107

Revelation
1:9, 125
1:16, 126
2:12, 126
2:16, 126
6:1–17, 126–128
6:8, 126
6:10, 126
7:1, 129
9:7–11, 128–129
9:13–21, 129
9:16, 129
12:7–8, 129–130
13:3, 125
13:7–8, 130–131
13:10, 126
13:14, 126
16:12, 131
17:9, 125
17:9–14, 133–135
17:15–18:24, 131–133
17:16, 132
17:16–17, 133
17:18, 131
18:3, 132
18:9, 132
18:9–10, 132
18:11–20, 133
18:16, 128
18:23, 133

19:11–16, 135
19:11–21, 133–135
19:13–14, 128, 135
19:15, 126
19:19, 135
19:21, 126

Rabbinic Literature

b. Avodah Zarah
62a, 8

m. Avodah Zarah
5.6, 43n14

Avot of Rabbi Nathan
27b, 8

y. Bava Qamma
4.4a, 28, 43n 19

t. Betzah
2.6, 27, 43n14

Eruvin
3.5, 43n13

Lamentations Rabbah
1:52, 82

Leviticus Rabbah
30:6, 44n31

y. Nedarim
4.9.38d, 43n19

b. Shabbat
33b, 34, 45n38
121a, 43n19
145b, 82

t. Shabbat
13.9, 43n19

y. Shabbat
15d, 43n19

Midrash Sifre Devei Rav
317, 47n52

Targum Neofiti Genesis
37:36, 99n17
40:3–4, 99n17
41:10–13, 99n17

Targum Pseudo-Jonathan
Numbers
34:15, 17n9

b. Yoma
35b, 8

y. Yoma
8.5.45b, 43n19

**Greek and
Roman Literature**

Achilles Tatius
Leucippe and Clitophon
7.7–16, 86

Acts of Andrew
54, 97

Aelian
Nature of Animals
4.1, 121

Ambrose
Death of Theodosius
3.6, 101n31

Appian
Civil Wars
5.132, 117

Apuleius
Metamorphoses
9.39, 67
11.14, 123n21

Aristides
Orations
26, 133

Artemidorus
Onirocritica
33, 123n14

Assumption of Moses
6–7, 26

Aulus Gellius
Attic Nights
10.3.1–20, 42n10

Aurelius Victor
Book of the Caesars
41.4, 97

Cassius Dio
55.26.4, 117
56.19.1–5, 44n29
56.29.3, 101n31
57.4.3–4, 117
60.17.5–6, 87
69.12.2, 36
73.8.4, 8

Cicero
Letters to Atticus
6.5, 131

Republic
6.21–22, 101n31

Roscius the Comedian
28, 8

Against Verres
2.5.161–162, 42n10

1 Clement
37, 85
61, 85

Dead Sea Scrolls
1QM (War Scroll), 26,
52, 56
4Q405, 56
4Q503, 56

Demades
Orations
1.65, 123n20

Digest
1.5.4, 115
48.20.6, 99n17

Diogenes Laertius
4.64, 101n31
6.12–13, 114

Dionysius of Halicarnassus
Roman Antiquities
2.56, 101n31

Diogenes of Sinope
Cynic Epistles
10, 114

1 Enoch
Passim, 56
56:5–8, 129

2 Enoch
67.1–2, 101n31

Epictetus
Diatribes
3.22.69, 123n21
4.16.14, 114

Epiphanius
Panarion
3.27, 114

Eusebius
Ecclesiastical History
5.5, 84
8.12.3, 43n13

Preparation for the Gospel
8.14, 101n31

Firmicus Maternus
Mathesis
8.26.6, 99n17

Florus
Epitome
1.1, 101n31

Gospel of Peter
4.14, 97

Hesiod
Theogony
319–324, 136n10

Historia Augusta
Hadrian
10.2, 122
10.4, 122

Homer
Iliad
2.529, 135
2.830, 135
6.181–182, 136n10
23.706–739, 107

Horace
Odes
3.26, 123n21

Ignatius
Polycarp
6, 85

Joseph and Aseneth
8.2, 102n37
17.5, 102n37

Josephus
Against Apion
2.202, 102n41

Jewish Antiquities
3.80, 110
9.55, 110
11.325–339, 111
14.20, 43n20
14.195, 43n20
14.309, 101n31

14.408, 10
14.417, 18n23
15.111–146, 17n10
15.296, 8, 16n1, 18n24
15.353, 8
15.409, 102n43
16.12–15, 111
16.163, 102n37
16.285, 4
16.292, 4
16.399, 93
17.23–31, 17n9
17.25–26, 5
17.167, 101n31
17.198, 5
17.266, 21n45
17.282–283, 35
17.299, 90
18.60–62, 34
18.109–115, 58
18.136–137, 58
18.269–278, 28
19.257, 117
19.357, 103n50
19.364–366, 84
19.366, 12
20.100, xx
20.105–112, 82
20.108, 27
20.115, 27
20.119, 28
20.122, 16n1
20.132, 21n45
20.136, 21n45
20.169–171, 26, 88
20.176, 4, 8, 16n1,
    18n20

Jewish War
1.45, 93
1.210, 5
1.249, 93
1.282, 102n43
1.299, 10
1.308, 18n23
1.366–384, 17n10

1.403, 8, 16n1, 18n24
1.527, 21n45
1.535, 21n45
1.672, 5
2.52, 16n1, 17n10,
    21n45
2.56, 5
2.58, 16n1, 21n45
2.63, 14, 21n46
2.74, 16n1
2.175–177, 34
2.192, 28
2.220, 21n49
2.224, 27
2.224–227, 82
2.229, 27
2.233, 28
2.236, 16n1
2.261–263, 26, 88
2.298–300, 14
2.319, 21n45
2.429, 93
2.430–437, 17n6
2.244–246, 21n45
2.450–454, 21n45
2.460, 27
2.477, 27
2.490–497, 21n49
2.494, 13
2.500, 21n48, 105n81
2.503–506, 13
2.508–509, 13
2.646, 13
3.12, 17n7
3.35–36, 5
3.66, 16n1
3.68, 21n48, 105n81
4.442, 28
5.367, 58
5.371, 58
5.412, 58
5.550–566, 13
5.563, 106n97
6.127, 58
6.300–309, 58
6.348, 58

7.26–36, 73
7.47–52, 14–15,
    21n50
7.217, 37

Life
33, 21n45
35, 15, 90
37, 15, 90
242, 20n42
403, 93

Juvenal
Satire
16, 77

Lactantius
Divine Institutes
4.26, 97

Life of Adam and Eve
38.2, 56
40, 56

Livy
28.19, 115
28.34, 70
31.39, 126

Martyrdom of Dasius
Passim, 84

Martyrdom of Julius the
    Veteran
1.4, 84

Martyrdom of Marcellus
1.1, 84

Marinus
Proclus
37, 101n31

Martial
Epigrams
6.64, 131

Mucinius Felix
36, 123n21

Origen
*Commentary Series on the
Gospel of Matthew*
140, 98

Ovid
Amores
1.9, 123n21

Art of Love
3.233, 123n21

Festivals
485–498, 101n31

Tristia
4.7.13, 136,10
5.7.13–15, 136n5

*Periplus of the Erythraean
Sea*
19, 31, 44n31

Petronius
Satyricon
57, 87

Philo of Alexandria
Confusion of Languages
34.174, 56

Flaccus
78–80, 42n10

Legum Allegoriae
3.82, 102n37

Legation to Gaius
157, 102n37
317, 102n37

Providence
2.46, 63

Plato
Apology
28E, 116, 123n21

Plautus
Poenulus
886, 97

Truculentus
638, 97

Pliny the Elder
Natural History
2.30, 101n31
9.60–65, 72

Plutarch
Aemilius Paullus
18.5, 126

Antonius
34.3–5, 136n5

Crassus
24.5–25.5, 136n5

Pelopidas
295a, 101n31

Romulus
27, 101n31

Polybius
1.80.13, 97
6.39.12, 77
21.38.1–6, 71

Propertius
1.6.30, 123n21
4.1.137, 123n21

Psalms of Solomon
17, 26

Pseudo-Lucian
Ass
27, 123n20

(Pseudo-)Quintillian
Declamations
6.9, 97–98

Seneca
Benefits
3.25

Epistles
107.9, 123n21
113.9, 136n10
120.12, 121

Ira
1.18.4, 99n17

Lucilus
109.8–9, 114

Marcia
9, 7
23–24, 7

Seneca the Elder
Controversies
9 pref. 4, 114

Sextus Empiricus
Against the Mathematicians
1.295, 123n20

Sibylline Oracles
1–5, 26
2.214–237, 56
5.399, 59

Strabo
5.3.7, 117

Suetonius
Augustus
27.4, 117
45.4, 42n10

Gaius Caligula
40, 117

Grammarians
13, 87

Nero
19.2, 117

Testament of Adam
3.6, 101n31

Testament of Asher
5.4, 102n37

Testament of Judah
22.3, 110

Testament of Levi
8.15, 110

Tacitus
Annals
1.17, 8
1.30, 117
2.16, 117
4.4, 18n19
13.7, 21n48
14.44, 121
15.1–19, 136n5
15.28, 21n49
16.5, 117
20.3, 117

Dialogue on Oratory
34, 114
37, 114

Germania
43, 100n26

Histories
1.11, 21n49
3.33, 132
3.47, 16n4
3.49, 21n54

Tertullian
Apology
5, 84

Crown
12, 84

Idolatry
17.2, 84

Vergil
Aeneid
6.782–783, 131

Georgics
1.463–468, 101n31
2.535, 131

Vettius Valens
5.2, 123n21
7.3, 123n21

Papyri and Inscriptions

AE
1932.27, 10
1976.653, 67
1983.380, 45n35
2004.1913, 21n43
2005.1737, 21n43
2010.1852, 21n43
2014.1641, 21n43

BGU
4, 43n21
69, 77
81, 44n30
140, 20n34
423, 15, 17n14
515, 31, 44n30
908, 43n21
1564, 19n32

CIL
13.1997, 123n14
16.103, 21n43

CPL
106, 45n40
304, 113

IG
2.2.3441, 106n95
4.2.122, 110
7.2713, 101n35

IGR
1.1337, 17n14
3.286, 101n35
4.201, 101n35
4.309, 101n35
4.310, 101n35
4.311, 101n35
4.314, 101n35
4.594, 101n35
4.1302, 101n35

IGRR
1121, 39–40
1122, 39–40

ILS
2413, 40
2658, 21n54

Mas
795, 102n43
796, 102n43
800, 102n43
801, 102n43
804, 102n43
805, 102n43
806, 102n43
807, 102n43
808, 102n43
809, 102n43
810, 102n43
811, 102n43
812, 102n43
813, 102n43
814, 102n43
815, 102n43
816, 102n43
817, 102n43
818, 102n43
821, 102n43
822, 102n43
823, 102n43

824, 102n43
825, 102n43
826, 102n43
850, 102n43
946, 102n43
947, 102n43
948, 102n43
949, 102n43
950, 102n43

O.Amst.
8, 43n23
9, 43n23
10, 43n23
11, 43n23
12, 43n23
13, 43n23
14, 43n23

O.Ber.
2.126, 36, 141

OGIS
266, 77
419, 106n95
420, 106n95
424, 106n95
496, 31

P.Amh.
2.77, 44n30

P.Corn.
90, 44n30

P.Dura
56, 35
97, 35

P.Fouad
8, 100n27

P.Gen.
1.17, 44n30

P.Lond.
23, 77
131, 8

P.Mich.
3.203, 40
7.422, 40, 48n68
7.425, 44n30
8.476, 18n16, 40,
  100n26
8.477, 18n16, 100n26
8.478, 18n16, 100n26
8.479, 18n16, 100n26
8.480, 18n16, 100n26
8.481, 18n16, 100n26
8.484, 40
10.582, 44n30

P.Oxy.
240, 76
285, 45n40
1022, 6, 17n14
1185, 44n30
1269, 45n40
1666, 40
2234, 29, 43n21

P.Strassb.
103, 77

P.Tebt.
1.48, 110
2.304, 27

P.Yadin
19, 38, 47n54, 47n57
20, 47n57

PSI
3.222, 31, 44n30

RGZM
56, 21n43

RMD
1.74, 21n43
5.348, 21n43

RMR
1, 17n14
87, 17n14

SB
5238, 27
5280, 43n21
9203, 44n30
9207, 28–29, 76
15496, 43n17
15497, 43n17
15498, 43n17
15499, 43n17
15500, 43n17

SEG
39.1711, 115

T.Vindol.
154, 10
257, 28, 43n18
281, 28, 43n18
302, 9
322, 28, 43n18
344, 28, 43n18

DMIPERP

§3, 21n45
§4, 14, 20n42
§5, 40, 52
§6, 52
§7, 52
§8, 52
§9, 27, 28, 42n11, 78
§10, 28, 42n11
§12, 13, 14, 20n42, 90
§13, 14, 67, 106n95
§14, 14, 62, 90
§15, 14
§16, 14
§17, 14, 62, 90
§18, 14, 62, 90
§20, 19n32
§21, 14
§22, 8, 35, 45n40, 46n47,
  72–73, 113–114, 135
§23, 5
§26, 13
§27, 44n26

§28, 14, 44n26
§29, 44n26
§30, 5, 14, 15, 21n47, 90
§31, 5, 14, 15, 21n47
§32, 5, 14, 15, 17n16,
    21n47, 62
§33, 17n13, 40, 52
§34, 5
§35, 14, 40
§36, 38, 46n47
§38, 14, 20n42, 21n47, 67
§40, 52
§41, 40, 52
§43, 38, 47n57
§49, 51, 52
§51, 62
§53, 40
§54, 40, 47n52
§55, 40, 47n52
§58, 38
§59, 44n26
§60, 13
§61, 52
§67, 18n20
§69, 36
§71, 6
§75, 36, 78
§83, 9, 13
§84, 9, 13
§85, 9, 13, 18n20
§86, 9, 13, 18n20
§87, 9, 13, 18n20
§88, 9, 13
§89, 9, 13
§90, 9, 13
§91, 9, 13
§117, 42n9
§118, 42n9
§119, 42n9, 103n50
§120, 5, 6, 10, 42n9,
    103n50
§121, 34, 35
§122, 34, 35
§123, 34, 35
§124, 34, 35
§125, 34, 35
§126, 34, 35

§127, 34, 35
§128, 34, 35
§129, 34, 35
§130, 34, 35
§136, 17n7
§145, 5, 15
§147, 17n13, 17n15, 40, 84
§148, 21n47, 40, 47n52
§149, 35
§158, 17
§160, 12, 28
§161, 17n7
§162, 17n13
§167, 6, 15, 17n16, 135
§168, 6, 15, 17n16, 135
§169, 6
§170, 6
§171, 6
§172, 6
§173, 6
§174, 6
§175, 6
§176, 6
§177, 6
§178, 6
§179, 6
§183, 6
§184, 5, 17n13
§189, 73
§190, 36
§192, 5
§193, 5
§194, 48n65
§195, 48n65
§197, 90
§198, 17n13, 40
§199, 5, 40
§200, 5, 14, 40
§201, 90
§202, 12
§203, 12
§204, 12, 17n13, 40
§205, 12
§206, 12
§207, 12
§208, 11, 12
§209, 12, 17n13, 40

§210, 17n6
§211, 17n6, 47n54
§212, 17n6
§213, 17n6
§214, 17n6
§215, 17n6
§216, 17n6
§217, 17n6
§218, 17n6
§219, 17n6
§220, 17n6
§221, 17n6, 17n8
§222, 17n6
§223, 17n6
§224, 17n6
§225, 17n6, 17n7, 84
§226, 17n6, 17n7, 84
§227, 17n6, 17n7, 84
§229, 17n6, 17n7, 84
§230, 17n6, 84
§231, 17n6, 84
§232, 17n6
§233, 17n6
§234, 17n6
§235, 17n6
§236, 17n6
§237, 17n6
§238, 17n6
§239, 17n6
§240, 17n6
§241, 17n6
§242, 17n6, 17n7
§243, 17n6
§244, 17n6
§245, 17n6
§246, 17n7
§247, 17n7
§248, 17n7
§249, 17n7
§250, 17n7
§251, 17n7
§252, 17n7
§253, 17n7
§254, 17n7
§255, 17n7
§256, 17n7
§257, 6, 17n7, 40

§258, 17n7
§259, 17n7
§263, 40
§293, 6, 40
§294, 5, 6, 17n13, 40
§295, 17n13, 17n15, 40, 92
§296, 5, 6, 40
§297, 34
§302, 34
§303, 34
§304, 34
§305, 34
§306, 34
§307, 34
§308, 34
§309, 34
§310, 34
§311, 34
§312, 34
§313, 34

§314, 34
§315, 34, 67
§336, 34
§337, 34
§338, 34
§339, 34
§340, 77
§342, 84, 143n2
§343, 84, 143n2
§344, 84, 143n2
§345, 84, 143n2
§346, 84, 143n2
§347, 84, 143n2
§348, 84, 143n2
§349, 84, 143n2
§350, 84, 143n2
§351, 84, 143n2
§352, 84, 143n2
§353, 84, 143n2
§354, 84, 143n2

§355, 84, 143n2
§356, 84, 143n2
§357, 84, 143n2
§358, 84, 143n2
§359, 84, 143n2
§360, 84, 143n2
§361, 84, 143n2
§362, 84, 143n2
§363, 17n13
§364, 6, 21n49
§365, 6, 21n49
§366, 6, 21n49
§367, 6, 21n49
§368, 6, 21n49
§369, 6, 21n49
§370, 6, 21n49
§371, 6, 21n49
§372, 6, 21n49
§373, 6, 21n49
§382, 111

About the Author

Christopher B. Zeichmann received his Ph.D from St. Michael's College in the University of Toronto and teaches at Emmanuel College in the University of Toronto and Ryerson University. He is editor of the forthcoming book *Essential Essays for the Study of the Military in New Testament Palestine* with Wipf and Stock Press. His research has been published in *Catholic Biblical Quarterly, Method and Theory in the Study of Religion, New Testament Studies, Early Christianity, Journal for the Study of the Historical Jesus, The Bible and Critical Theory*, various other journals, and several collected volumes. He also created and runs the Database of Military Inscriptions and Papyri of Early Roman Palestine [ArmyofRomanPalestine.com].

Lightning Source UK Ltd.
Milton Keynes UK
UKHW012256080119
334959UK00009B/192/P